The Illustrated Directory of

TRACTORS

Fordson
EJT 139
Fordson

The Illustrated Directory of

TRACTORS

GREENWICH
EDITIONS

A Salamander Book

Published in 2004 by
Greenwich Editions
The Chrysalis Building
Bramley Road
London W10 6SP
United Kingdom

An imprint of Chrysalis Books Group plc

ISBN 0-86288-687-2

All correspondence concerning the content of this volume should be addressed to Salamander Books Ltd.

Credits

Project Manager: Ray Bonds
Designers: Interprep Ltd
Reproduction: Anorax Imaging Ltd
Printed in Italy

The Author

Peter Henshaw has enjoyed a life-long interest in tractors, and has written several books about them, as well as motorcycles. A freelance writer and journalist, he was formerly Editor of *Motorcycle Sport & Leisure* magazine.

Acknowledgements

Thanks are due to the many tractor restorers and enthusiasts who allowed their machines to be photographed for this book. Many were pictured at the Dorset Steam Fair (England), the Rollag Steam Show (Ohio), and Mount Pleasant Labor Day Weekend. Also to Trevor Cuff and Ben Welsh, whose generous loan of tractor books and magazines was an invaluable help. And to Anna, for her patience through all those long hours and missed weekends. Without any of them, the book would never have got beyond the ideas stage.

This book is for Alexander, who loves tractors.

Peter Henshaw,
Sherborne, Dorset, February 2002.

Contents

Challenger
CAT

Introduction

The 20th century saw an agricultural revolution. It transformed not just yields per acre, but the way farmers worked, and their relationship to the land. And the cause of all this was the tractor. As the century dawned, steam power was already in use on farms, but only the big ones. Steam traction engines were complicated, expensive machines, needing specialist operators.

But gasoline or kerosene tractors, once Henry Ford had worked his cost-cutting magic, offered a true mechanical horse at a fraction of the price. Unlike a horse, the tractor didn't get tired, and didn't need feeding when there was no work to do. On the other hand, just a like a horse, the tractor of the day was on a more human scale than the massive traction engines. Those little Fordsons and Grey Fergies became familiar friends, almost part of the family. How else do you explain the intense affection for these old machines?

But they weren't bought for sentimental reasons. Tractors had to do a job of work, and as the years went by they got better at it. Early tractors were still relatively expensive, but Henry Ford changed all that with the affordable Fordson F. The Fordson was an all-round tractor, but it still couldn't cultivate row-crops. That came with the International Farmall of 1924, a tractor that really could do it all. Allis-Chalmers pioneered pneumatic rubber tyres in the 1930s, which dramatically increased efficiency, comfort and speed. Meanwhile, Harry Ferguson's three-point hitch revolutionised the tractor – 70 years later, his hydraulic implement control is still the industry standard.

The Depression had seen many smaller manufacturers go bust in both the

USA and Europe, but the survivors forged ahead. After World War II, more innovations came thick and fast: multi-speed transmissions, diesel engines and turbocharging; electronics; and air conditioning. But although each year appeared to bring bigger, more sophisticated tractors, there was still a market for simple, straightforward models. In fact, that's what this book is about, to show the sheer variety of tractors that have been and still are being produced, used and loved all over the world.

The Nebraska tests mentioned in the text refer to tractor tests undertaken at the University of Nebraska, USA. From 1920, these instigated standardised tests for tractor performance, and they remain the industry benchmark. For any tractor historian, they are invaluable. Some tractor terminology may need explaining though. Two power figures are usually given: PTO or belt power refers to the power take-off (for driving implements), .the equivalent of power at the crankshaft. Drawbar power equates to what the tractor can pull – ie, power at the rear wheels. Also, readers on either side of the Atlantic should take note of the difference between gasoline/petrol, kerosene/paraffin and distillate/TVO.

Now, exactly 100 years since those first tractors spluttered their way into the fields, the industry faces new challenges. Modern tractors need to be more efficient and less polluting than ever before; they'll need to use less fuel and cause less damage. The next 100 years should be even more interesting.

Below: Oliver's supercharged Super 99, the most powerful tractor of its time.

Advance-Rumely Type E

USA

Engine: Oil-cooled, two-cylinder
Bore x stroke: 10in x 12in (250 x 300mm)
Capacity: 1,884ci (29,390cc)
PTO power: 75hp
Drawbar power: 50hp
Transmission: Two-speed

Advance-Rumely tractors were unique: the great, stately Oil Pulls, with their huge two-cylinder kerosene engines and quirky engineering. The Rumely company had been founded in 1853, and built a variety of threshers and steam engines. In the early years of the 20th century, Dr Edward Rumeley (grandson of the founder) decided to build a tractor that could run on cheaper, heavier fuels like kerosene. These needed a higher compression than gasoline, and were prone to pre-ignition.

Rumely's answer was to cool the new tractor's twin cylinder engine with oil, which would allow the engine to run hotter. Better still, an oil cooling system wouldn't rust or freeze up. And to prevent pre-ignition, a Secor-Higgins carburettor (named after the Oil Pull's designers) atomised water along with the fuel. It was unusual, but it worked. The first Oil Pulls started work in 1910, and Rumeley went on building them on the same basic format until 1931, when the company was taken over by Allis-Chalmers.

The Type E pictured here was the biggest offered by Rumeley, rated at 30 drawbar horsepower and 60 at the belt – tested at Nebraska, it measured 50hp/75hp, staggeringly powerful for the time. Oil Pulls could work as field tractors, but they were favoured for belt work, thanks to their heavy, smooth running engines.

Below: Advance-Rumely Oil Pulls owed much to heavy steam engines.

Advance-Rumeley Type H 1918

USA

Engine: Oil-cooled, two-cylinder
Capacity: 654ci (10,202cc)
PTO power: 30hp
Drawbar power: 16hp
Transmission: Two-speed
Weight: 9,506lb (4,278kg)

Advance-Rumeley Oil Pulls were made in a whole range of sizes. The first one of 1910, the B Model, was rated at 25/45hp; there were also the E (30/60hp, 1910-23); the F (18/35hp, 1911-18); the G (20/40hp, 1918-24); the H (16/30hp, 1917-24); and the K (12/20hp, 1918-24). All of these used a variation on the Oil Pull twin-cylinder motor apart from the F, which used a single-cylinder based on

components from the big E. The E was the biggest tractor of its time, a fact which Rumeley demonstrated in graphic fashion: they hitched three of these monsters to a 50-row plough, which worked an acre every four and a half minutes, and ploughed 2,000 acres over six days.

The H pictured here started out as a 14/28, but was soon uprated to 16/30. It also brought in some significant changes. It used a 180-degree crankshaft, instead of the 360-degree crank used previously. This smoothed out the power pulses and allowed the use of a smaller crankshaft and flywheel for the better balanced motor. The H's cylinders were cast in pairs (another new feature) and its valves were incorporated in the cylinder-head. Ignition used low-tension switches, rather than high-tension spark plugs. This was also the most popular of the Oil Pulls, but despite a good following among threshermen and sawmills they were eventually overcome by smaller, more nimble tractors.

Below: Advance-Rumely's H was in production for seven years.

AGCO 9630

USA

Specifications for LT85 (2002)
Engine: Water-cooled, four-cylinder
Bore x stroke: 4.0 x 4.7in (102 x 120mm)
Capacity: 236ci (3.9 litres)
PTO power: 85hp @ 2,200rpm
Transmission: 24-speed
Weight: 8,150lb (3,300kg)

The AGCO story is extraordinary. Within a few years, the company grew from a management buy-out of the US arm of Deutz-Allis to be America's largest tractor manufacturer, with plants in Britain and France as well as the USA.

The story really began when Allis-Chalmers sold its tractor business to Deutz of Germany, which closed down Allis production in 1985. But four years later it restarted, with Deutz-Allis tractors using Deutz air-cooled engines in American chassis. A management buy-out soon followed, and AGCO was born.

At first, the same Deutz-Allis tractors continued, but now under the AGCO-Allis name. Massey-Ferguson and White-New Idea were taken over by AGCO in 1993, followed by McConnell in 1995.

Clearly, this was no cut-price, asset-stripping takeover, but a long-term strategic plan to stay in the tractor market, with big money behind it. AGCO had a simple strategy of making use of all the famous names it now owned, but sharing components between three different tractor lines.

The 9000 series (9630 pictured here) illustrated that well, with a choice of Deutz air-cooled or Cummins water-cooled diesels, and the latest electronics developed by AGCO as a whole. Specifications refer to the LT85, part of the new RT, DT and LT ranges of 2001.

Below: AGCO-Allis 9630 shown here was a hybrid of American and German technology. AGCO went on to develop all-American tractors.

AGCO 8630

USA

Specifications for RT145 (2002)
Engine: Water-cooled, six-cylinder, turbo-diesel
Bore x stroke: 4.02 x 4.72in (102 x 120mm)
Capacity: 359ci (5.9 litres)
PTO power: 145hp @ 2,200rpm
Transmission: 18-speed (32-speed optional)
Weight: 16,894lb (7,662kg)

By the mid-1990s, AGCO had a full range of tractors on offer, many of them new or uprated. The smallest 5650/5660 used a 3.0 litre three-cylinder air-cooled diesel (part of the Deutz legacy) in 45 or 55hp form.The 5600 had the choice of 63 or 72hp from a four-cylinder 4.0 litre diesel, while the 6600 was a row-crop version of the same thing – these two line-ups shared 72 per cent of their components. The 6600 did have a higher power option though, in the 81hp 6690.

The 7600 series took farmers into the 100hp+ range. There was a basic 7600 with an 89hp five-cylinder engine (still an air-cooled Deutz) but the 7630 used a 115hp six and the 7650 was turbocharged to give 128hp. The standard transmission on all three was a 24-speed forward, 12-speed reverse set-up, though a 30-speed system was optional on the 7650. Hydrostatic steering, two-speed PTO and air conditoning all featured as well.

Next step up was the 8600 – an 8630 is shown here. This had no more power than the 7600 (a choice of 103 or 120hp) but was more of a high-tech tractor. AGCO had certainly embraced the electronic age, and one innovation on the 8600 was an electronic speed control. All the driver had to do was set the required speed, and the tractor would maintain it regardless of load.

Below: With the big 8600 series, AGCO introduced electronic control.

AGCO 9670

USA

Specifications for DT 225 (2002)
Engine: Water-cooled, six-cylinder, turbo-diesel
Bore x stroke: 4.59 x 5.35in (117 x 136mm)
Capacity: 505ci (8.3 litres)
PTO power: 225hp @ 2,200rpm
Transmission: 18-speed
Weight: 19,700lb (8,935kg)

With the 9600 series, AGCO-Allis made several pointers to the future. First, it offered water-cooled diesels of 7.6 and 8.7 litres, underlining that the Deutz air-cooled legacy was on the way out. Actually, the highest powered 9600 did use an air-cooled 9.6 litre six of nearly 200hp, but not for long – by 1996, the Deutz engines had been phased out altogether.

Secondly, the level of electronic control seen in the 9600 (9670 seen here) showed the way tractors were going. Electronic control – of engines,

transmissions and implements – allowed more efficient, accurate work.

Not that it reduced the driver to an unskilled steerer. He or she was now bombarded with information about engine, ground and PTO speed, the amount of acres worked and tyre slippage. And the driver still had to set the job parameters to start with. That done, the AGCO-Allis Electronic Control system would in theory monitor the three-point hitch, rear differential lock, four-wheel-drive, PTO transmission and hydraulic pressure, adjusting any of them as necessary.

The top 9600 was the 9655, with four-wheel-drive, and 155hp at the PTO from its 466ci six-cylinder turbo diesel. In 1999, the latest update was the 9755, now with 200 PTO hp. It was part of a new four-model high power range, from 160 to 225hp. All of them used AGCO-Allis' own 1300 series turbo diesels. It looked like the Deutz legacy had finally been laid to rest.

Below: AGCO's biggest tractor of the mid-1990s, the 9600 series.

Allis-Chalmers

USA

Allis-Chalmers turned to making tractors only after plunging into bankruptcy in 1912. Founded in the mid-19th century, the company was involved in iron, mining, steam and electricity, but General Otto Falk, put in to turn Allis-Chalmers around, knew an expanding market when he saw one. The strategy worked, and before long A-C was prospering. Its great leap forward came with the Model U and pneumatic rubber tyres. Until then, all tractors had used plain steel wheels, which limited road speeds to a fast walking pace – the U could reach a heady 15mph on the blacktop.

Other innovations followed, such as powershift adjustable track and

hydraulic lift with the WD tractor. But turbocharging, which arrived with the D19 diesel in 1961, was a real milestone in tractor technology. Soon afterwards, A-C was first again, with the market's first 100hp machine, the D21. The company was an early user of hydrostatic front-wheel-drive as well. But A-C tractors couldn't survive the tough 1980s on their own, and were taken over by Deutz, which in turn was taken over by the giant AGCO concern in 1990. And so the Allis name has survived, as part of the AGCO empire.

Below: Allis-Chalmers made a huge variety of tractors from 1914.

Allis-Chalmers 20-35

USA

Specifications for 1921 18-30
Engine: Water-cooled, four-cylinder
Capacity: 461ci (7,192cc)
PTO power: 38.6hp
Drawbar power: 23.6hp
Fuel consumption: 9.81hp/hr per gallon
Transmission: Two-speed
Gear speeds: 2.6mph and 3.2mph (4.1 and 5.2km/h)
Weight: 6,640lb (3,018kg)

The 20-35 was Allis-Chalmers' big tractor of the 1920s and '30s, an update of the 18-30 which could trace its roots back to the 15-30 of 1918. A-C was still a relative newcomer to the tractor market, and this showed in the design of both 20-35 and earlier 18-30 – they were dependable machines, but heavier and more expensive than mass – produced rivals such as the Fordson. In 1924, Allis sold about 600 20-35s – by the following year, Henry

Ford had shifted over one *million* Fordsons.

Word was slowly getting around that A-C engineering, not to mention its 461ci (7.2 litres) four-cylinder motor, was tough and reliable, but at over $2,000 it cost too much. George Gardner, then manager of A-C's Tractor Divison, resigned his post and advised company boss General Falk to close the whole thing down. He didn't. Instead, he promoted assistant manager Harry Merrit into the hot seat: Merrit was a hands-on operator, full of energy, and immediately began a serious cost cutting exercise. A 20-35 was taken apart, piece by piece, and every piece scrutinized – if it could be made simpler and cheaper, or just thrown away, it was.

The effect was dramatic. In the same year, the 20-35's price was slashed to $1,885. The following year it was cut again, to $1,495, and yet again the year after (1928) to $1,295. By 1934, it was being listed at $970, less than half the original price! Sales rose, and A-C survived.

Below: Massive price cuts helped the 20-35 survive the Depression.

Allis-Chalmers Model WC

USA

Specifications for 1934, pneumatic tyres
Engine: Water-cooled, four-cylinder
Capacity: 202ci (3,151cc)
PTO power: 21.5hp
Drawbar power: n/a
Fuel consumption: 10.17hp/hr per gallon
Transmission: Four-speed
Gear speeds: 2.5mph-9.25mph (4.0-14.9km/h)
Weight: 3,792lb (1,723kg)

By 1931, Allis-Chalmers was poised on the brink of greatness. The innovation of pneumatic rubber tyres with the Model U had brought it wide publicity, and the purchase of Advance-Rumeley in 1931 gave access to a nationwide spread of dealers across the USA. But it still didn't have a mass-market tractor to take advantage of this – the WC model changed all that.

Launched in 1933, it wasn't wildly innovative, but did come along at just the right time to take advantage of the post-Depression boom in two-plough tractor

Below: The WC was one of the most successful Allis-Chalmers machines.

demand. Only 3,000 were sold in 1934, but over 10,000 the year after and 29,000 in the peak year of 1937. In fact, the WC carried on selling well right up until it was dropped from the range in 1948.

There were reasons for this success. A-C still had a head start over its rivals in the provision of rubber tyres, which had now been around long enough to overcome the trade's initial conservatism. They cost more to buy (the rubber WC cost $150 more than the steel wheeled version) but Nebraska tests showed that they were far more fuel efficient: the WC gave 5.62hp/hours per gallon of fuel on steel, 8.18 on rubber. Not only that, but rubber was quieter, allowed higher speeds and easier steering, and didn't hurt off-road traction either.

But it wasn't just rubber tyres that accounted for the WC's success. Its

Allis-Chalmers Model A

USA

Specification for 1936
Engine: Water-cooled, four-cylinder
Capacity: 461ci (7,192cc)
PTO power: 44hp
Drawbar power: 33hp
Fuel consumption: n/a
Gear speeds: n/a
Weight: 7,425lb (3,375kg)

A-C's Model A was too little, too late. While the company's smaller U and WC models took the tractor market by storm (largely thanks to their pneumatic tyres) the A was intended as a direct replacement for the heavyweight Model E, the 18-30/20-35 series. As such, it did incorporate some useful updates on the A. The old two-speed transmission was replaced by a new four-speeder (really a beefed up version of the smaller U's gearbox) and there was a pneumatic tyre option of course. In fact, the engine was the only major component that remained unchanged. Using petrol, it had a 4.75in (119mm) bore and produced 33hp at the drawbar, 44hp at the belt. If you opted for distillate, the engine was supplied slightly larger, with a 5-inch (125mm) bore. A-C later boosted the power to nearly 40hp drawbar, and over 50hp at the belt.

The trouble was, by the time the A was unveiled in 1936, the tractor market was changing. With its large powerful engine, the A was designed more for powering threshers than towing implements. But in the late 1930s, the first combines began to trickle onto the market – self-powered, these didn't need tractors to tow them. Oddly, A-C itself seems to have overlooked the A's *raison d'etre* – the same year it announced the A, it ceased production of Advance-Rumeley threshers (now under the Allis wing) which

201ci (3.2 litres) four-cylinder motor wasn't especially big, but was lighter than traditional power units, and the first in a tractor to feature 'square' dimensions – nothing to do with its attitude to rock n' roll, more that the bore and stroke measurements were the same, at four inches! Making the chassis of high tensile steel saved more weight, so despite modest power ot 21hp at the drawbar, the WC was able to perform as well as more powerful tractors. A drive-in attachment for implements was another useful feature, which combined with the optional power lift arm to give a poor man's Ferguson system.

Restyled in 1938 along the modern lines of little brother B model, the WC went on selling well, and 178,000 were sold in all. It was one of the most successful tractors A-C ever made.

---

were a major reason for the A's existence.

At 3.75 tons, the A was a big, heavy and thirsty beast – it was said that the smaller U could do nearly as much work on half the amount of fuel. It was, in short, a bit of a dinosaur. Sales reflected this, and even in its best year, fewer than 600 found homes. When it was dropped after five years (a short life for an A-C machine) a mere 1,200 had left the works.

Below: In some ways, the Model A was outdated even when new.

Allis-Chalmers Model B

USA

Specifications for 1938
Engine: Water-cooled, four-cylinder
Capacity: 116ci (1,810cc)
PTO power: 14.0hp
Drawbar power: 10.3hp
Fuel consumption: 11.14hp/hr per gallon
Gear speeds: 2.5mph-7.75mph (4.0-12.5km/h)
Weight: 2,260lb (1,027kg)

Allis-Chalmers had more than its share of landmark tractors, and the little Model B was one of them. As late as the mid-1930s, many smaller farms in America continued to rely on horse teams, or even human power – tractor prices had come down over the years, but they were still out of reach of farmers like these. For them, Allis unveiled the Model B in 1937. Smaller, lighter and cheaper than almost any other tractor then available, it opened up a whole new market, and was a great success. Eleven thousand were sold in the first year alone.

The secret of the B's success was its price – it cost a mere $495 brand new – partly thanks to its simple, light construction. Allis returned to unit construction with the B, using the engine, transmission and torque tube as stressed members which absorbed loads and helped support each other and the rest of the tractor. They also allowed a clean, wasp-waisted look that was capitalised on by the famous industrial designer Brooks Stevens. Allis

commissioned him to style the B, and he produced a rounded-off design quite distinct from other lumpy, square-rigged tractors of the day. On a more practical note, the narrow waist allowed the driver good visibility of the work in hand, while retaining a centrally mounted engine.

The engine itself was a small one – the first batch of 96 Bs used a 113ci (1.8 litres) Waukesha unit, but A-C soon came up with its own 116ci (1,81 litres) four-cylinder engine, which produced 10.3hp at the bar, 14 at the belt, according to Nebraska tests. It wasn't much, but Allis did increase the capacity to 125ci (2.0 litres) in 1943, which gave a useful power boost.

As well as being an affordable tractor, the Model B could easily be upgraded: a PTO shaft and belt pulley, for example, cost an extra $35, an adjustable-width front axle, just $20 (allowing ten different spacings from 38 to 60 inches. And recognising that small farmers' needs were just as varied as big ones, A-C offered a whole range: the IB was the industrial tractor ("drives like a car" said A-C when it was announced in 1939); the Asparagus Special was a high-clearance machine; and the Potato Special a narrow track. Rubber tyres were standard on all of these, except in the depths of wartime, when a rubber shortage forced A-C temporarily back to solid steel wheels.

By the time production ended in 1957, over 100,000 Bs had been sold, confirming Allis-Chalmers' role as a major US tractor manufacturer.

Below: Less than $500, brand new, the Allis B was a lesson in economy.

Allis-Chalmers Model U

USA

Specifications for 1935, pneumatic tyres
Engine: Water-cooled, four-cylinder
Capacity: 301ci (4,697cc)
PTO power: 31hp
Drawbar power: 22.7hp
Fuel consumption: 9.94hp/hr per gallon
Speeds: 2.3-10mph (3.7-16.1km/h)
Weight: 5,140lb (2,336kg)

The U was a real breakthrough for Allis-Chalmers, thanks to one dramatic innovation – pneumatic rubber tyres. Until then, all agricultural tractors ran on steel rims, which were tough, but limited speed on the road to a good walking pace. Some industrial tractors had solid rubber tyres, which allowed higher speeds on road but had little traction off it. An agricultural tractor needed speed in both places.

The story goes that two A-C dealers in Iowa were converting tractors to run on truck tyres. A company rep got wind of this, and reported back to Harry Merrit, who got to work with characteristic en-thusiasm. He obtained a couple of aircraft tyres from Firestone, while suitable rims were made up. The first results were disappointing, until the recommended pressure of 70psi was reduced to 12psi. That made a dramatic improvement. The new tyres halved the U's power requirements in the field, which allowed a high-speed fourth gear to be added to the trans-mission – A-C claimed 5mph (8km/h) while ploughing, 15mph (24km/h) on the blacktop. Extra speed meant less working time, lower costs and higher profits.

As if that wasn't enough, the new tyres made the tractor quieter, easier to steer and more com-fortable to ride on. Traction was not compromised either, and fuel economy was far better. Still, the company wisely started off by making the new-fangled pneumatics a $150 option, rather than bumping up the price of the whole tractor. The conservative farming community still needed convincing, so A-C fitted extra high-speed gearing to some machines, hired some big-name drivers and toured the county fairs with a tractor race! An air-equipped U even reached 68mph (109km/h) on the Utah salt flats.

Oddly, the tractor which pioneered all this, the U, nearly never happened at all. United Tractors & Farm Equipment (a Chicago-based co-operative) asked A-C to develop a 4,000lb (1,818kg) machine it could sell under its own name. The co-op folded,but A-C had the tractor out only a year later, at first with a Continental side-valve engine, later with A-C's own UM, slightly bigger and with an extra 4hp. Whatever, the U was very long-lived: in 23 years of production (1929-52) well over 20,000 were built.

Below: Revolution! The first production tractor with pneumatic tyres.

Above: Pneumatic tyres improved efficiency, and cut working time.

A-C held demonstration races of the U, at 60mph+!

Allis-Chalmers Model M

USA

Specifications for 1935
Engine: Water-cooled, four-cylinder
Capacity: 301ci (4,697cc)
PTO power: 31.6hp
Drawbar power: 22.8hp
Fuel consumption: 9.79hp/hr per gallon
Speeds: 1.8-4.2mph (2.9-6.8km/h)
Weight: 6,855lb (3,116kg)

The Model M was purely and simply a crawler version of the successful U, but naturally enough without the rubber tyres! So it used the same 301ci (4.7 litre) four-cylinder as the U, rated at 22.8 drawbar hp and 31.6 at the belt. However, the gearing was very much lower than that of the agricultural tractor, a fact underlined when the M was tested at Nebraska in August 1935. In low gear it pulled just over 5,000lb (2,273kg), or 73 per cent of its own weight – that was at a speed of 2.05mph (3.3km/h) while developing 27.4hp. In second gear, the pull was reduced to 2,700lb (1,227kg), albeit at 3.13mph (5.0km/h). After 50 hours of testing, recorded by the thorough men of Nebraska, no repairs or adjustments were needed.

Interesting to note though, that only a couple of weeks after the Allis-Chalmers M had made its trip to Nebraska, a much bigger crawler from Caterpillar was tested. Significantly, this had a diesel engine, a sign of the times for agricultural tractors as well as crawlers. A-C crawlers would be using General Motors diesel engines in the 1950s.

Allis-Chalmers Model WF

USA

Specifications for 1934 WC, pneumatic tyres
Engine: Water-cooled, four-cylinder
Capacity: 202ci (3,151cc)
PTO power: 21.5hp
Drawbar power: n/a
Fuel economy: 10.17hp/hr per gallon
Speeds: 2.5-9.3mph (4.0-15.0km/h)
Weight: 3,792lb (1,724kg)

Commencing production in 1940, the WF was no more nor less than a standard-tread version of the row-crop WC. Both tractors used A-C's own W-type engine with a four-inch (100mm) bore and stroke, rated at 1,300rpm. However, the war intervened, and a lack of materials affected production of the WF – no electric starter and lights were fitted in 1941 (from No 1903) and the model ceased production altogether in 1943. It resumed the following year, before stopping for good in 1951. For 1948, it was still available with steel wheels, and listed at $971 – with pneumatics, you paid $1,210, which now really did include the electric starter and lights.

Unusually for a US-made tractor, the WF was never tested at Nebraska. Again, the war was at least partly to blame. All tractor testing ceased in November 1941, and didn't resume again until '46. So we don't have any specific figures for the WF, but as it was mechanically identical to the WC, whose specifications can be used as a guide.

Right: The Allis-Chalmers WF was mechanically identical to the WC.

Above: Model M crawler could pull nearly three-quarters its own weight.

Allis-Chalmers Model G USA

Specifications for 1948
Engine: Water-cooled, four-cylinder
Capacity: 62ci (967cc)
PTO power: 10.9hp @ 1,800rpm
Drawbar power: 9hp @ 1,800rpm
Transmission: Four-speed

Allis-Chalmers' Model G was radical not just by its own standards, but those of the entire tractor industry. Until 1948, when the G was unveiled, every tractor had followed the traditional layout of engine up front and driver behind. The G was rear engined, its little 62ci (967cc) four-cylinder motor balanced just behind the rear wheels. This put 80 per cent of the tractor's weight over its rear wheels, which was obviously a great help to traction. Just as important, it gave the driver a completely unrestricted view of the work going on below, a particular boon when working on smaller, more delicate crops.

In fact, the G Model itself was small, smaller than any other machine then available – this was not a tractor for the Mid-Western prairies. Instead, it was aimed at nurseries and vegetable gardens, and at those farmers who couldn't even afford the bargain basement Model B, which was A-C's smallest 'real' tractor. Allis didn't make an engine small enough to suit the G, so bought in a 62ci (967cc) four from Continental, which at 1,800rpm gave 9 drawbar hp and 10.9hp at the belt. Some customers wanted more power, asking for a 20hp G, but A-C never made one, perhaps afraid that it would steal sales from the evergreen B.

Allis-Chalmers Model WD45 USA

Specifications for 1953
Engine: Water-cooled, four-cylinder
Capacity: 226ci (3,526cc)
PTO power: 40.47hp
Drawbar power: 30.18hp
Fuel economy: 10.64hp/hr per gallon
Speeds: 2.5-11.3mph (4.0-14.7km/h)
Weight: 3,955lb (1,798kg)s

This was Allis-Chalmers' conventional mid-range tractor, announced in 1948 to replace the WC. But while it looked conventional (with just silightly more power from the same 201ci / 3.1 litres engine, thanks to increased compression) the WD was brimming with new ideas.

Power Shift wheels took all the drudgery out of changing the rear tread width. Until now, one had to laboriously jack up the rear of the tractor and knock the wheels into place with a sledgehammer. Power Shift used the tractor's own power to move the wheels in or out on spiral rails, a simple but effective system which ranked alongside other milestones of tractor technology like the three-point hitch. It is still in use today. Two-Clutch Power Control allowed continuous power take-off, whether the tractor was moving or not, and the oil-bath transmission clutch could take any amount of slippage. Traction Booster was A-C's take on hydraulic implement control (present on most tractors by this time). A cam-driven hydraulic pump on the mainshaft between engine and transmission clutches would lift the implement just enough when the going got sticky, to prevent slippage and keep

Above: Rear-engined, Model G was the smallest tractor on sale in 1948.

But even with its modest power, the G wasn't badly equipped, with a four-speed gearbox (with first low enough to allow 0.75mph / 1.2km/h at part throttle) and adjustable track between 36 and 64 inches (900-1,600mm). A hydraulic lift was optional,at $99, a belt pulley cost $19 extra. In keeping with its radical design, the G wasn't made at the old West Allis works, but at a newer factory in Gadsden, Alabama – it was even assembled in Dieppe, in France, as well. It ceased production in 1955.

Right: The WD45 bristled with innovations like in-seat hitching.

the tractor moving.

But buyers wanted more power as well. The WD45 update had a bigger engine (thanks to an extra half-inch of stroke) plus something called 'Power Crater'. The latter was a concave piston crown that allegedly increased turbulence. It worked, too, as Nebraska tests showed the WD45 to have 25 per cent more power than the plain WD. The new Snap Coupler allowed genuine in-seat hitching of implements, and the new WD45D was A-C's first-ever diesel tractor. With no suitable diesel of its own (the crawlers were already using GM units) Allis bought up the Buda company of Harvey, Illinois, which built a six-cylinder diesel of about the right size. The WD45, in both petrol and diesel form, lasted until 1957, and underlined the 1950s as a time of innovation in tractor design.

Allis-Chalmers D14

USA

Specifications for 1957, petrol
Engine: Water-cooled, four-cylinder
Capacity: 149ci (2,324cc)
PTO power: 32.6hp
Drawbar power: 24.5hp
Fuel consumption: 12.01hp/hr per gallon
Speeds: (high range) 2.2-12.0mph (3.5-19.2km/h)
Weight: 3,623lb (1,647kg)

In 1957-59 the D-series range – the D10, 12, 14 and 17 – replaced A-C's entire line-up of B, CA, WD and really the long-dead E and A as well. It was a big and varied range, with over 50 models available once the little D10/12 came on stream.Lke everything else in the modern world, tractors were becoming more specialised, and Allis had to keep up or fall by the wayside.

The D14 pictured here was the mid-range three-plough machine, which partly replaced the long-running WD. It had an all-new Allis-designed engine, a 149ci (2.3 litres) four, though it was derived from WD experience, with Power Crater pistons. At 1,650rpm, it gave 34hp at the belt – there was an LPG option as well. Perhaps bigger news was a transmission innovation this tractor shared with its bigger brother D17, which was launched at the same time: Power Director.

At the time, most tractors made do with four transmission speeds, but Power Director effectively doubled these to eight, by providing a choice of high and low ranges. A second clutch span at 70 per cent of engine speed via two reduction gears. it ran in an oil bath (just like the old WD's hand clutch) and was

Allis-Chalmers D12

USA

Specifications for 1959, petrol
Engine: Water-cooled, four-cylinder
Capacity: 139ci (2,168cc)
PTO power: 28.6hp
Drawbar power: 23.6hp
Fuel consumption: 10.73hp/hr per gallon
Speeds: (high range) 2.0-11.4mph (3.2-18.2km/h)
Weight: 2,945lb (1,339kg)

Just as the D14 and D17 replaced the WD, the D10 and 12 did the same for the little Model B, the tractor cheap enough for small family farms. Sales of the B had actually dwindled to the point where it was dropped in 1957, two years before its replacements were ready. They were mechanically similar, though the D10 was the single-row version, the D12 the two-row. Both used a smaller 139ci (2.2 litres) version of the D14's engine, with Power Crater pistons. Tests of a D10 at Nebraska produced 28.5 PTO hp and 25.7 at the drawbar.

As with the B, they could be had with most of the features of A-C's bigger tractors, such as Traction Booster, Power Shift rear wheels and independent PTO. What they didn't get was the multi-speed Power Director transmission. However, in 1961, the D10 and 12 did benefit from the 14's 149ci (2.3 litres) engine, which meant a 17 per cent power boost. What they never had was a diesel option, perhaps one reason why they were never very successful – in nearly a decade, their combined sales were a little over 9,000.

However, there was one interesting experiment with the D12. In 1959 Allis-Chalmers unveiled a fuel cell powered version. With 1,008 cells fuelled by a

Above: Two-speed ranges gave eight forward speeds to the D14.

tough enough to be used as a variable speed device, which some owners did. It also enabled high-low shifting on the move around the eight forward and two reverse speeds.

The D14 soon got more power, becoming D15 in 1960 with an 18 per cent boost, thanks to higher compression and a 2,000rpm rated speed. And for the first time, there was a diesel option, a four-cylinder version of the D17's six.

Above: The D12 was A-C's smallest tractor of the early 1960s.

mixture of propane and other gases, and a 15kw electric motor, the fuel cell D12 was pollution-free, silent and relatively efficient. Unfortunately, it also weighed over 5,000lb (2,272kg) and produced the equivalent of 20hp. Back to the drawing board, then.

Allis-Chalmers D272

USA (UK)

Engine: Water-cooled, four-cylinder
Powe: 30hp
Transmission: Four-speed

To American tractor enthusiasts, the Allis D272 is something of a rarity – it was never made by A-C in America at all. The company had set up an assembly plant in Southampton in the south of England, which began building Model Bs from 1948. At first, this was a pure assembly operation, but Britain's tractor market was expanding, and the UK offshoot soon moved to a bigger factory in Lincolnshire.

Allis UK was also moving beyond being a mere assembly operation, redesigning the B to suit British conditions. In 1955, two years before the B was dropped by A-C in America, it came up with its own update, the D270. There were three significant changes: the addition of a live PTO (which was needed for work with a Roto-Baler); a four-speed gearbox replaced the three-speeder; and reflecting the increasing trend towards diesel power, there was a diesel option. Perkins provided the diesel, the P3/143 three-cylinder. It had actually offered more power than the original petrol version, though cost an extra $145. The D272 pictured here followed in 1959, with another power boost (30hp from petrol) and improved hydraulics.

In 1960, the English arm of A-C also announced the bigger ED-40 which ran alongside the D272. This used a 138ci (2.2 litres) Standard-Ricardo diesel of 38hp

Allis-Chalmers D19/D21

USA

Specifications for D19
Engine: Water-cooled, six-cylinder, turbo
Capacity: 262ci (4,087cc)
PTO power: 67hp
Drawbar power: 62hp

The 1960s saw a horsepower race in the American tractor market. Once, 45hp had been the 'big' class. Then it was 60, 70, 80 and so on. Among the smaller machines, it was quite straightforward to increase the size of existing engines, or even just boost compression and rated speed without changing anything else. But for bigger tractors, things got more complicated. Two Allis machines of the early '60s – the D19 and D21 – showed two alternative approaches.

Originally, the D19 was intended as a modest 60hp step-up form the D17, but it became clear that this wouldn't be enough. So there was what some called 'an eleventh hour programme' to beef it up. The 262ci (4.1 litres) diesel could have been enlarged to 290ci (4.5 litres), until someone pointed out that a turbocharger would give the same power boost, and be quicker to develop. It worked, liberating an extra 25 per cent power, to give 67 PTO hp and 62hp at the drawbar. There was a 70hp gasoline version as well – for 1961, these were big figures.

With the D21, which followed a couple of years later, Allis followed a different path. After the success of the D19, it seemed likely that the company would opt to turbocharge its first 100hp machine as well. In the event, it came up with a massive new direct-injection diesel engine which relied on cubic inches rather than a turbo, to get results. With 426ci (6.6 litres) and 103hp, it again took A-C into new markets, and so big was the D21 that it needed an all-new transmission and new range of implements to suit – a 7-bottom plough

Above: English Perpendicular. D272 was an English adaptation of the Model B.

(later 41) and was exported to Canada. But sales were disappointing, and in 1968 A-C sold its Lincolnshire factory,which spelt the end for both ED-40 and D272.

Above: One hundred horsepower! The square-rigged D21 power house.

was among them. It was also A-C's first diesel-only tractor, reflecting the market's slow but sure conversion to the advantages of this form of fuel. The D21 was in production for six years, by which time it had been turbocharged as well. The power race waited for no man.

Allis-Chalmers One Seventy USA

Specifications for 1967, diesel
Engine: Water-cooled, four-cylinder
Capacity: 236ci (3,682cc)
PTO power: 54hp
Drawbar power: 39.4hp
Fuel consumption: 16.38hp/hr per gallon
Transmission: Eight-speed
Gear speeds: 2.0mph-13.3mph (3.2-21.4km/h)
Weight: 5,950lb (2,705kg)

In the 1960s, A-C gradually replaced its entire range of D-series tractors with the new squared-off look of the 100-series. The One Ninety was first in 1965, replacing the well regarded D19. With it came a whole new family of petrol, LPG and direct injection diesel engines – these shared many components, but made up a range of six units, with four or six cylinders. A turbocharged One Ninety soon followed, when it was apparent that the standard 301ci (4.7 litres) wasn't powerful enough to keep up with 90hp rivals.

Just as the new D-series style had started at the top and worked down in the 1950s, so the big One Ninety was followed in 1967 by the One Seventy and One Eighty, which replaced the D17. Although they looked new, these tractors actually mixed and matched many existing components – the Power Director, Traction Booster and D17 four-speed gearbox were all carried over unchanged. There was a new cab option though, as Nebraska was starting to voice concern about the increasing number of roll-over accidents.

Allis-Chalmers 4W305 USA

Specifications for 7580 (1976)
Engine: Water-cooled, six-cylinder, turbo/intercooled
Capacity: 426ci (6,646cc)
PTO power: 186hp
Drawbar power: 127hp
Fuel consumption: 14.81hp/hr per gallon
Transmission: 20-speed
Gear speeds: 1.5mph-17.6mph (2.4-28km/h)
Weight: 23,520lb (10,691kg)

Allis-Chalmers was a relative latecomer to the four-wheel-drive market, but in the 1970s and early '80s it produced several all-wheel-drive tractors. The theory was simple: tractors were getting more powerful by the year, with turbochargers added, then intercoolers, to ever larger diesel engines. But these were producing more power than two driven wheels could cope with, leading to traction problems – driving the front wheels as well was the obvious solution. A-C had actually built a few four-wheel-drivers in the early 1960s. The T16 Sugar Babe (intended for use in sugar plantations) was really an Allis-engined version of Deerfield's existing TL16 loader. This giant proved not reliable enough for non-stop field work, and most ended up working with sugar cane. It had four- or six-speed transmission options, and Allis' own 344ci (5.3 litres) turbocharged diesel. There were other attempts to design an in-house 4x4, but when these failed A-C simply bought the proven Steiger Bearcat and renamed it the 440. About 1,000 of these were sold between 1972 and '76. All were powered by a Cummins V8 of 555ci (8.7 litres) driving through a 10-speed transmission.

Right: Modernist, squared-off 100-series replaced the D-models.

The One Seventy came in petrol or diesel form, the latter using one of A-C's first bought-in engines for quite a while. It was from Perkins, the well known English make, a four-cylinder 236ci (3.7 litres) unit which produced just over 50 PTO hp, 39hp drawbar. The petrol version, which used A-C's own 226ci (3.5 litres) four, offered slightly more PTO power, according to Nebraska, slightly less drawbar, and of course it was thirstier. The One Eighty used an uprated version of this engine, but serious vibration problems meant it was soon dropped. These two tractors were uprated into the 175/185 a couple of years later, and the diesels hung on right up to 1981.

Above: Nearly the end of the line, Allis-Chalmers' 4W305

Finally, in 1976, A-C replaced the 440 with a four-wheel-drive tractor of its own design. The 7580 was very much a 4x4 version of the existing 7000-series, using the same 426ci (6.1 litres) six-cylinder diesel with turbo and intercooling. It was rated at 186hp at the PTO, and also used the 7080's 20-speed transmission and disc brakes. An even bigger 8550 used the massive 844ci (13.1 litres) engine from one of A-C's crawlers. It had twin turbos (though no intercooler) and produced 254hp, according to Nebraska tests.

These were replaced in 1982 with what proved to be part of A-C's swansong. The 4W220 and 4W305 (pictured on previous page) featured slightly derated engines compared to the 7580, but with the latest 8000-series cab. The 220 used A-C's own 670-HI diesel (426ci/6.1 litres turbo intercooled) and the 305 the 731ci 6120T.

Right: Four-wheel-drive, articulation, twin wheels – the modern 4x4 tractor.

Allis-Chalmers 7000 Series USA

Specifications for 7020 (1977)
Engine: Water-cooled, six-cylinder, turbo/intercooled
Capacity: 301ci (4,697cc)
PTO power: 124hp
Drawbar power: 102hp
Fuel consumption: 13.1hp/hr per gallon
Transmission: 12-speed
Gear speeds: 1.9mph-19mph (3.0-30km/h)
Weight: 15,610lb (7,095kg)

By the early 1970s, the Allis-Chalmers range was looking decidedly old-fashioned. There had been some significant updates, but everyone offered turbos now, and the eight-speed Power Director transmission had long since been eclipsed by 12- and 16-speed rivals. So when the 7000-series 'Power Squadron' was launched in Los Angeles in 1973, it was to some no doubt delighted and relieved A-C dealers. The 7000 (with its distinctive forward-sloping nose) would characterise A-C right through the '70s.

The whole range used A-C's familiar 426ci (6.1 litres) diesel, now in 130hp turbo and 156hp intercooled forms. Power Director was still there, but now uprated to 20 forward speeds. The hydraulic system was load-sensive, in that pressure and flow automatically adjusted to the work in hand – another Allis first. Just as significant for the driver was the Acousta-Cab, which A-C claimed was the quietest in the business – the company

Right: The 7000 series was A-C's big hope for the 1970s.

had already pioneered a quieter cab with the 200 in 1970, which was isolated from the chassis altogether.

An entry-level tractor, the 7000, was unveiled in 1975, to replace the 200. It actually carried over 200 parts, though was uprated to 106hp, had the Acousta Cab and a 12-speed power shift transmission. But it was another four years before the cheapest 7000-series machine became a genuine smaller version of the 7030. Over 36,000 7000-series A-Cs were built in the 1970s.

Austin

UK

Engine: Water-cooled, four-cylinder
Power: 26.5hp
Transmission: Two-speed (French model, 3-speed)
Weight: 3,136lb (1,411kg)

Sir Herbert Austin – England's Henry Ford – announced his tractor in 1919, two years after Henry's Fordson astonished the world with its light weight and low price. But he had been dabbling in tractors before then, selling the American-made Interstate Plowman in Britiain during WWÎ. He also shipped over a Denning tractor, with a view to ordering 300 in the first year, but a problem with the test model put him off.

But it didn't put him off farm machinery altogether. A three-wheel 20hp

prototype, the Culti-Trac, never made production, but the simple, conventional four-cylinder Austin tractor did. It used the 20hp engine from an Austin car (though rated higher), but with a different sump to allow for unit construction. The new machine did well at the 1919 Lincoln Tractor Trials, and Austin built a factory in France, as well as producing it in Birmingham. But, as with so many contemporaries, it was simply too expensive compared to a Fordson – about three times the price, to be exact. From 1927, all production was transferred to France, where the Austin survived longer thanks to tariff barriers. It was actually reintroduced to Britain (as a French import) in 1931, with a bigger engine and 3/4 plough capability, to no avail.

Below: English Fordson? But the Austin failed to emulate Henry's success.

Avery Model A

USA

Specifications for 45/65 (1920)
Engine: Water-cooled,
Bore x stroke: 7.75 x 8.0in (194 x 200mm)
Capacity: 1,509ci (23,537cc)
PTO Power: 69hp
Drawbar Power: 50hp @ 634rpm
Transmission: Two-speed
Speeds: 2 and 3mph (3.2 and 4.8km/h)
Fuel consumption: 9.06 gallons per hr
Weight: 22,000lb (9,900kg)

Avery's first tractor wasn't really a tractor at all, more of a hybrid.Its first internal combustion engined device was the Model A of 1909. It was intended as a tractor which could travel at a reasonable road speed for transport work. Top speed was 15mph, with the four-cylinder engine producing 36hp at 1,000rpm.

That was assuming you went for the solid rubber tyre option. The alternative was steel wheels with small indentations around the surface, into which hardwood pegs could be hammered for maximum traction. Avery' s transport tractor-truck was an interesting concept, but perhaps before its time. Now of course, on-road performance is a key factor in machines like the JCB Fastrac.
Avery had more success when it tackled the pure tractor market, with a series of flat-twin and flat-four engines. The first of these was the 20-35 twin of 1912. Not big enough? The following year, they coupled two of these together to create the 40-80. This was the largest tractor Avery ever built – its flat-four displaced 1,507ci and the whole machine weighed 9,900kg. It was indeed a monster, but had to be renamed the 45/65 when Nebraska tests revealed that it couldn't make that claimed 80 belt hp.

Below: Avery later concentrated on small tractors like this one.

Avery 12/25

USA

Specifications for 12/25 (1921)
Engine: Water-cooled,twin-cylinder
Bore x stroke: 6.5 x 7.0in (163 x 175mm)
Capacity: 464ci (7,238cc)
PTO power: 25.2hp @ 700rpm
Drawbar power: 13.8hp @ 700rpm
Fuel consumption: 8.1hp/hr per gallon
Weight: 7,500lb (3,375kg)

Like many of its contemporaries, Avery had been in the agricultural supply business well before it started making tractors. The company had been founded in 1874, to make corn planters, though it later moved to Peoria, Illinois' and went on to produce steam-powered machines.

The early 1920s saw a whole range of Avery tractors and cultivators, most of them powered by variations on the water-cooled flat-twin engine. The 12/25

pictured here sat in the middle of the range, with its 464ci unit, and unusually it was fitted with a cab. Not that it aquitted itself very well when submitted for the standard University of Nebraska tests in March 1931. In three months' testing, the Avery managed only 35 hours of running time. The engineer's notes at the time give a clue as to why this might have been: 'Before any official data was taken, the rear exahust valve was ground and both spark plugs were replaced by long-skirted plugs. The pistons were taken out and turned down about 0.005inch. A tthe end of the maximum drawbar test, the compression of the rear cylinder was weak. 'Still, the 12/25 went on to better its rated power, with 13.8hp at the drawbar, 25.2hp at the brake.

The company was too small to stay independent for long. One of Avery's last tractors was the Hercules-powered Rotrack of 1939, which could be converted from standard to tricycle layout in less than 30 minutes.

Below: Twin-cylinder 12/25 Avery had a cab to keep the driver dry!

Big Bud 525/50 Diesel

USA

Engine: Water-cooled, six-cylinder, turbo intercooled
Bore x stroke: 6.25 x 6.25in (156 x 156mm)
Capacity: 1,150ci (17,940cc)
PTO Power: Not measured
Drawbar Power: 406hp @ 2,100rpm
Transmission: Nine-speed
Speeds: Not measured
Fuel consumption: 14.9hp/hr per gallon
Weight: 51,920lb (23,600kg)

Today, massive 500 horsepower tractors are nothing very unusual, especially in North America, where big fields and bigger farms make these monsters economically viable. It made sense, then, that the pioneer of these super-tractors was American. The odd thing was, it wasn't Case-International, or Allis or John Deere, but a small independent. The Big Bud, unveiled in 1968, was bigger than any other field machine of the time, starting off with a 250bhp six-cylinder diesel, and ending up with 525bhp by '77. With its massive bonnet,

high and wide cab, full rounded mudguards and equal-sized wheels, it looked more like a mutant, highly modified truck than a tractor. In truth, it was a cross between the two, with truck-based power plants but tractor-type transmission and four-wheel-drive. It also had a removable Power Train Skid, so that the complete motor could be removed for servicing.

For the second series of Big Buds, announced in 1977, the cab was widened to 60 inches (1,500mm), with power ratings from 320 to 525bhp. As with the first series, the cab was tiliting to facilitiate engine access, and in fact much of this Big Bud was the same as the original.

A third series went on sale in 1979, now with power options up to 650bhp and mostly using Twin Disk Power Shift transmissions (which a few of the second series had been built with). A ROPS safety cab was later added, before the fourth series Big Bud arrived in 1986. This had a new oscillation system and used mainly 12-speed power shift transmissions. Power was up to 740bhp. More recently, the Big Bud 16V-747 boasted a V16 Detroit Diesel two-stroke, with twin turbochargers and 760bhp at 2,100rpm.

Left: Mr Big – Big Bud and Steiger pioneered a new breed of super-tractor.

Bristol 20

UK

No specifications data available

Bristol, as you might expect, was based in Bristol, England. It was actually an offshoot of Roadless Traction, the company that built countless half-track conversions on Fordson and other tractors. Bristol made the rubber jointed tracks that Roadless used in its conversions.

But none of the adapted tractors that Roadless produced really fulfilled the role of a true, purpose-built small crawler. This Bristol arranged itself.

The result was similar to the Ransome MG crawler, with full tracks and tiller steering, though later models used the more conventional lever controls. The first Bristol was built in 1933 by Douglas, the motorcycle manufacturer, and

early Bristols actually used a Douglas engine, the air-cooled 76.9ci (1,200cc) horizontally-opposed twin. Douglas motorcycles of the time were similar in layout to twin-cylinder BMWs, but smaller. Following interest in the venture by the Jowett car company, a Jowett water-cooled 7hp engine (another opposed twin) was fitted, and later a Coventry Victor 10hp diesel was added as well. In either case, the transmission was three-speed, with a drawbar pull of 2,000lb (900kg). Bristol was later bought by HA Sanders, and after WWII Austin 10 and A70 gasoline car engines were used. Bristol was eventually taken over by Marshall in the early 1970s.

Below: Simple is efficient. Bristol made crawlers for 40 years.

Bolinder-Munktell

SWEDEN

Specifications for Bolinder-Munktell 320 (1961)
Engine: Water-cooled, three-cylinder diesel
Power: 40hp
Transmission: Five-speed
Weight: 1,570kg

This Swedish tractor maker was the result of a merger between manufacturer Munktell and its engine supplier, Bolinder, in 1932. Munktell had been a pioneer tractor builder in Sweden, starting with the BM 30-40 in 1913. Right from the start it used Bolinder engines. The company experimented with wood burning motors, as Sweden imported all of its oil, and in the early 1920s offered a two-

stroke twin-cylinder machine, with ignition by hot-bulb. The latter was an interesting system, derived from a marine engine, and used compressed air to start after the hot-bulbs had been heated by an integral blowtorch.

This Model 22 was joined by the larger 20-30 in 1930, and a 31hp machine (still with twin-cylinder hot-bulb motor) in 1939. In fact, it wasn't until the 1950s that Bolinder finally abandoned the hot-bulb, unveiling the BM35 and BM55 with conventional direct injection diesel engines. Bolinder-Munktell later merged with truck, bus and car maker Volvo, to form Volvo BM, whose tractor division was in turn was taken over by Finnish tractor maker, Valmet.

Below: Swedish Bolinder-Munktell was later taken over by Volvo.

Case

USA

Some tractor companies are consistently innovative – others reluctant to take any sort of risk. The JI Case Threshing Company was arguably both, though not at the same time. As a maker of agricultural steam engines (they started in 1863) Case was well established, well respected, conservative. But it began experimenting with the new gasoline engine very early, building a gas tractor in 1894. That one didn't work, but once the company finally started making gasoline tractors commercially in 1912, it introduced eight new designs in its first seven years. These included the famous Crossmotor, which mounted the engine across the chassis – this allowed for a more rigid, longer lasting design.

A great success, the Crossmotor tractors took Case through the 1920s, and were replaced by the equally popular L and C models in 1929. But in the '30s

Case began to fall behind – it was late and reluctant to bring out a competitor for the small Farmall, and next to streamlined machines like the Oliver 70, Case tractors looked old-fashioned. It wasn't until the late 1950s that Case finally caught up, with the all-new 400, a modern diesel engine and contemporary styling. The company never fell behind again, with new options like Case-o-Matic, four-wheel drive and the Comfort King line. Case was taken over by Tenneco Inc in 1970, which gave it the stability not only to survive the tough decades that followed but to buy up David Brown in 1972, International Harvester in 1985 and Steiger in '87. In 1999, the company merged with New Holland, itself owned by Fiat.

Below: Case was a tractor pioneer, still in business today.

Case 10-20 Crossmotor

USA

Specifications for Case 10-18
Engine: Watercooled, 4-cylinder
Bore x stroke: 3.875 x 5.00in (97 x 125mm)
Capacity: 236ci (3,682cc)
PTO power: 18.4hp @ 1,050rpm
Drawbar power: 11.2hp @ 1,050rpm
Transmission: Two-speed
Speeds: 2.25mph and 3.5mph (3.6 and 5.6km/h)
Fuel consumption: 6.25hp/hr per gallon
Weight: 3,760lb (1,692kg)

Early tractors had riveted chassis that flexed and bent, pulling transmission gears out of alignment. The answer, unveiled by Case in 1915, was to mount the engine crossways, allowing a more rigid chassis – the Case Crossmotor was born.

A suitable engine almost fell into their lap. Case had recently bought the Pierce Motor Company, which already made a suitable four-cylinder overhead valve motor. It was redesigned for tractor use, with the cylinders and upper half of the crankcase cast in one – the cylinder-head was removable though, and there were ports in the crankcase to allow the cleaning of coolant passages.

The 10-20 was a three-wheeler, using only a single driven wheel (which eliminated the need for a differential) though the second rear wheel could be clutched in for difficult going. Ready to roll, the 10-20 weighed around 5,000lb (2,250kg) and cost $900. It sold well, and was soon followed by the 9-18, a four-wheel tractor which developed the Crossmotor concept. Launched in 1916 with

Case Model L

USA

Engine: Water-cooled, four-cylinder
Bore x stroke: 4.625 x 6.00in
Capacity: 403ci (6,287cc)
PTO power: 45.0hp @ 1,100rpm
Drawbar power: 36.8hp @ 1,100rpm
Transmission: Three-speed
Speeds: 3.5-5.6mph (5.6-9.0km/h)
Fuel consumption: 10.05hp/hr per gallon
Weight: 8,025lb (3,611kg) (on rubber tyres)

If the 10-20 marked the beginning of the Crossmotor era, the L of 1929 marked its end. Instead of being mounted transversely, the four-cylinder engine sat longitudinally in the frame. It replaced both 18-32 and 25-45 Crossmotors, but marked the trend towards lighter weights by being the size and weight of the smaller model, with the capabilities of the big one. The motor itself was of 403ci (6,287cc), rated at 1,100rpm, sitting in a unit-construction frame – this effectively used the engine as part of the chassis structure, part of the reason for the L's lower weight.

Transmission was by three-speed gearbox via a roller chain final drive between the differential and rear axle – this last feature was designed by Case engineer David P. Davies, and patented. There was also a rear power take-off for using the new generation of shaft-powered binders and combines. The L was another hit for Case, and well over 30,000 were sold before it was finally dropped in 1940. Some farmers loved the low-revving John Deere D, but the Case L produced more power from its smaller higher revving engine, and used

***Above:* The innovative Crossmotor established Case in the tractor business.**

a steel fabricated chassis, it dropped this for a one-piece cast iron frame a couple of years later, though it retained the 236ci (3,682cc) four-cylinder engine. Five thousand were sold in its first three years, and Case stuck with the Crossmotor concept for the next decade.

***Above:* The Case Model L marked the end of the Crossmotor era.**

only a littel more fuel, at 6.52 horsepower hours per gallon of kerosene. The Deere also had only a two-speed gearbox and no PTO.

Although there were no major changes throughout the L's long production run, rubber tyres were offered from 1934 and the Model LI variant came with dual turning brakes. The industrial version also had an electric start option, which wasn't offered on the standard model until late in the production run. In 1940, the L was replaced by the LA, really a restyled updated version of the same tractor, with four-speed gearbox – that one stayed in production until 1953.

Case Model C

USA

Engine: Water-cooled, four-cylinder
Bore x stroke: 3.875 x 5.50in (97 x 138mm)
Capacity: 259ci (4,040cc)
PTO power: 29.8hp @ 1,100rpm
Drawbar power: 19.6hp @1,100rpm
Transmission: Three-speed
Speeds: 2.3-4.5mph (3.7-7.2km/h)
Fuel consumption: 11.36hp/hr per gallon
Weight: 4,155lb (1,870kg)

The Case Model C was in nearly every respect a scaled-down L. Its 324ci four cylinder engine had removable cylinder liners and a three-bearing crankshaft, just like the L. But, unlike the L, this three-plough Case was aimed directly at the Fordson. At the height of the 1920s tractor wars, which saw many American tractor makers go bust, Henry Ford was selling his machine at below cost price, a tough call for any rival to meet.

Compared to a Fordson, the Case C was heavier (around 30 per cent more, at over 4,000lb, 1,800kg) but less inclined to tip over backwards. It was slightly less powerful – tested at Nebraska in 1929, it yielded 29.8 belt hp and nearly 20 at the drawbar – but it was a lot more economical (11hp/hr per gallon against 5.9). But, as with every other rival, it cost a lot more to buy – one reason why Ford was selling around 2,000 machines a month in the late 1920s, while Case sold 300 or so Model Cs.

The CC which followed in 1930 was a general purpose version of the same

Case Model RC

USA

Specifications for Case RC (1936)
Engine: Water-cooled, four-cylinder
Capacity: 133ci (2,075cc)
PTO power: 17.6hp @ 1,425rpm
Drawbar power: 11.6hp @ 1,425rpm
Fuel consumption: 9.87hp/hr per gal
Gear speeds: 2.3-4.5mph (3.7-7.2km/h)
Weight: 3,350lb (1,508kg)

The Case RC – Case's stab at the small tractor market in 1935 – almost never made it to production at all. Despite urgent requests from dealers and salesmen, who saw the IH Farmall cleaning up in the sub-20hp market, president Leon Clausen was reluctant, afraid a new smaller Case would take sales from the more profitable CC model. In the end, he relented, but insisted there should be no launch fanfare, and even vetoed the addition of fourth gear to the transmission, which would have cost Case a mere 72 cents! Clausen was partially vindicated when the RC was not a great seller.

Its 133ci (2,075cc) four-cylinder engine, bought in from Waukesha, and rated at 1,425rpm, was a gasoline only motor.There was a conventional radiator and fan, but no water pump – the thermosyphon effect was found to be sufficient. As a single-plough general purpose machine, the RC had an adjustable rear track of 44 to 80 inches (1,100-2,000mm), achievable by a combination of sliding the wheels on splines and reversing them – an adjustable front track was among the options, as were rubber tyres. In 1938, a standard track version – the R – was introduced, otherwise mechanically identical to the

Above: Model C was aimed at the Fordson. Too expensive though.

machine, a natural competitor for the John Deere GP. With its crop clearance, rear PTO and adjustable rear wheel spacing, this was a definitive general purpose machine. A useful tractor, but customers wanted a smaller, lighter machine than the CC – the RC was Case's answer.

Above: Case needed a Farmall rival – the 18hp RC did the job.

RC. Initially, the RC used overhead steering, but there were complaints that the ratio was too high, so Case's characteristic 'chicken roost' side arm steering soon replaced it. Both R and RC got that 72-cent fourth gear in 1939, as well as the new 'sunburst' radiator grille and Flambeau Red colour. But that was the last gasp, and in that same year both R and RC were replaced by the S model.

Case Model S

USA

Specifications for Case S gasoline (1953)
Engine: Water-cooled, 4 cylinder
Bore x stroke: 3.625 x 4in (91 x 100mm)
Capacity: 165ci (2,574cc)
PTO power: 32hp @ 1,600rpm
Drawbar power: 28hp @ 1,600rpm
Transmission: 4-speed
Speeds: 2.5-10.3mph (4.0-16.5km/h)
Fuel consumption: 10.28hp/hr per gallon
Weight: 5,007lb (2,253kg)

The year 1939 was a significant one for Case. Flambeau Red, the Case trademark colour, made its first appearance, and a new generation of tractors took over in the next couple of years. The C was replaced by the D model; the LA replaced the L; the V was a new single-plough machine; and the S was a

two-plough general purpose tractor which replaced the R and RC.

In fact, the S was all-new from the start, with its own four-speed gearbox and a whole range of different versions. The S was the standard tread version; SC denoted the general purpose tractor; SO was designed for orchard work; and SI was the industrial model. There was also an SC-4 with a fixed tread wide front axle, but this was only built in 1953-54.

The S faced plenty of competition in the two-plough tractor field, as this and the three-plough machines were the biggest sellers in North America. John Deere offered the Models A (two-plough) and G (three-plough), while International Harvester was pushing the identical looking Farmall H and M, and Oliver the 70 and 80. There were few significant changes in the life of the S, apart from an engine capacity hike for the final year, the introduction of hydraulics and a change from hand to foot clutch.

Below: The S was Case's all-new tractor for 1939, in Flambeau Red.

Case Model DC4

USA

Specifications for Case Model DC
Engine: Water-cooled four-cylinder
Bore x stroke: 3.875 x 5.5in (91 x 138mm)
Capacity: 259ci (4,040cc)
PTO power: 35.5hp @ 1,100rpm
Drawbar power: 24.4hp @ 1,100rpm
Transmission: Three-speeds
Speeds: 2.5-5.0mph (4.0-8.0km/h)
Fuel consumption: 12.1hp/hr per gallon
Weight: 7,010lb (3,155kg)

Although basically a restyled C series, the Case Model D was upgraded over the years to keep it up to date. For 1940, a four-speed transmission replaced the three-speed, and later a foot clutch replaced the hand clutch, while the Motor-Lift was displaced in favour of hydraulics; disc brakes replaced the band type and, for the first time, Case offered an LPG (Liquid Petroleum Gas) conversion; a water pump did away with relying on the thermosyphon effect;

and a live PTO became standard. The DC4 shown here was the solid wide front axle version.

All the D series had swinging drawbars and belt pulleys, while rubber tyres were still optional, but it wouldn't be long before they became universal. The final change came in 1952, the penultimate year of production, when the Case Eagle Hitch was added. The Ferguson-Ford three-point hydraulic hitch had revolutionised tractor use: with in-built draft control, it made ploughing far easier and safer than before, while attaching and removing implements was simplicity itself.

It was such a success among farmers that all the major manufacturers had to come up with something comparable. Despite Case president Clausen's dismissal of the Ferguson device as 'a cheatin' system', Case had to follow suit, and the Eagle Hitch did indeed allow rapid snap-on fitting of implements, though it didn't have the Ferguson's draft control. It was all a matter of damage limitation however – by 1960, the draft-control three-point hitch had become virtually universal.

Below: Model D replaced the C in 1939, adding updates over the next 13 years.

Case Model DEX

USA

Specifications for Model DEX (1942)
Engine: Water-cooled four-cylinder
Bore x stroke: 3.875 x 5.5in (91 x 138mm)
Capacity: 259ci (4,040cc)
PTO power: 35.5hp @ 1,100rpm
Drawbar power: 24.4hp @ 1,100rpm
Transmission: Three-speeds
Speeds: 2.5-5.0mph (4.0-8.0km/h)
Fuel consumption: 12.1hp/hr per gallon
Weight: 7,010lb (3,155kg)

While the S model was Case's standard two-plough tractor, the D was the bigger three-plough version. It was also the first Case to use the new Flambeau Red colour scheme. 'Flambeau' came from the French for flame, or flaming torch – either that or the Flambeau River region of North Wisconsin, the original home of the eagle, Old Abe, which in turn was Case's trademark. Either way, it heralded a bright, bold new era for Case, its previous standard colour being a drab grey.

The Model D was also built in a huge array of different configurations. Like all its rivals, Case was learning that it was no longer sufficient to offer just one model as cheaply as possible. Farmers now wanted different tractors for different jobs: row-crop work, industrial, specialist crops such as sugar cane or vines, the list was endless. So D denoted standard-tread: DC3, DC4 and DH were all-purpose tractors, and the DC4 came as a Rice Special as well; DO was the orchard machine; DV for vineyards; there were two industrial tractors, the DI Standard and DI Narrow Tread; DCS sugar cane special;

Case Model SC

USA

Engine: Water-cooled, four-cylinder
Bore x stroke: 3.5 x 4.0in (88 x 100mm)
Capacity: 154ci (2,402cc)
PTO power: 21.6hp @ 1,550rpm
Drawbar power: 16.8hp @ 1,550rpm
Transmission: Four-speed
Speeds: 2.5-9.7mph (4.0-15.5km/h)
Fuel consumption: 10.55hp/hr per gallon
Weight: 4,200lb (1,890kg)

This was the Case company's two-plough tractor of the 1940s and early '50s, fighting for sales in a very competitive market sector against the likes of the Farmall H and Oliver 70. Announced in November 1940, the Case S was totally new, also sporting the striking Flambeau Red colour that all Case tractors used from 1939 to the mid-1950s. It came as basic Model S; the general purpose SC (pictured here); the SO orchard and SI industrial versions; and the SC-4, with a fixed tread wide axle that was only available for the last couple of years of production.

The engine, although petrol rather than diesel, was still a pointer to the future in using a relatively short four-inch stroke, and 3.5-inch bore. In 1953, the bore was increased to 3.625 inches and rated speed upped to 1,600rpm.

And, according to contemporary advertising, Case was in no doubt as to the SC's advantages: '22 New Conveniences,'" it trumpeted in a 1941 issue of *Successful Farming*. 'Synchronised steering with caster action stays free from play...deep-cushioned backrest seat...simplified 4-speed gearshift...electric

Above: Case DEX was the most popular of the Model Ds.

export versions of the D and DO were different again; and, finally, there were military versions of the DI during WWII. On top of all that, there were wheel options: single or twin fronts, rubber tyres or steel rims. Of all these, the basic D and all-purpose DCs were the best sellers, but the DEX pictured here was the most popular of the other variants: between 1940 and 1952 over 7,000 were built.

Above: '22 New Conveniences' on the Case SC, including electric starting.

starting...2-rate generator...self-sealed cooling pump, lubricated for life...built-in implement mounts...adjustable drawbar...No-one can tell you in words the feel of a good gun, the sensation of superb horse-flesh in the saddle, the thrill of driving this new SC Case tractor.'

Case Model VA

USA

Case: Model VAC
Engine: Water-cooled, four-cylinder
Bore x stroke: 3.25 x 3.75in (81 x 94mm)
Capacity: 124ci (1,940cc)
PTO power: 17.0hp @ 1,425rpm
Drawbar power: 12.5hp @ 1,425rpm
Transmission: Four-speed
Speeds: 2.3-8.4mph (3.7-13.4km/h)
Fuel consumption: 10.8hp/hr per gallon
Weight: 3,199lb (1,440kg)

Case boss Leon Clausen needed a lot of persuading. He was convinced that a small tractor would never sell, but in the late 1930s Case dealers could see little one-plough machines like the John Deere L and Allis-Chalmers B doing good business. Clausen eventually relented, and in 1940 the V-series was unveiled, using a Continental 124ci (1,934cc) motor and four-speed transmission. In typical Case fashion, it was both heavier and more powerful than rivals.

After a couple of years, it was replaced by the VA, which was part of the Case line up for the next decade or so. Much was new, such as the 124ci ohv engine, now built by Case itself at the Rock Island plant. It was also the first machine to use Eagle Hitch, Case's answer to the Ford Ferguson three-point hitch – it didn't have a Fergie's Draft Control, but allowed snap-on attachment of implements. There were lots of variations on the VA theme: VAC was the general purpose version; VAC-14 had a low seat with the driver straddling the

Case Model 500 (1952-1956)

USA

Engine: Water-cooled, six- cylinder
Bore x stroke: 4.0 x 5.0in (100 x 125mm)
Capacity: 377ci (5,881cc)
PTO power: 63.8hp @ 1,350rpm
Torque: 411lb ft (303Nm) @ 1,050rpm
Transmission: Four-speed
Speeds: 2.7-10.1mph (4.3-16.2km/h)
Fuel consumption: 15.8hp/hr per gallon
Weight: 8,128lb (3,658kg)

Case's first production diesel tractor, the 500 was basically a re-engined version of the LA, which itself could be traced back to the L Model of 1929. It still used the roller-chain final drive which the original LA had featured and retained its hand clutch – the final drive ratio was higher, but that and the engine were the only major changes.

The six-cylinder 377ci (5,881cc) diesel engine was all new, with seven main bearings (one between each connecting rod) which made it an immensely strong and durable motor – it was also one of the most powerful diesels on offer to American farmers in the 1950s. It was indirect injection (that is, the fuel was injected into the inlet tract, not directly into the cylinder) and used the Lavona 'power-cell' type combustion chamber, the same as used on the Oliver 88 and 99 machines, and by Mack trucks.

The 500 was no lightweight, tipping the scales at 7,500lb (3,375kg), and

Above: Despite Leon Clausen's reluctance, the little V was a success.

transmission, like the Ford-Ferguson; 15,000 VAI industrial tractors were built; VAS was a high-clearance row-crop machine, of which only 1,600 were made; and the VAH pictured here had adjustable front wheel spacing. Over 13 years, nearly 60,000 VA series tractors were turned out by Case.

Above: An all-new diesel engine marked out the Model 500.

working weight could approach 10,000. But it was a hard worker too – Nebraska tests showed that it could develop 7,400lb (3,330kg) of drawbar pull and 64hp at the belt. In 1956, it was replaced by the 600, which itself soon gave way to the 900. It may have been Case's first diesel, but the 500 was also the last Flambeau Red tractor.

Case 910B LPG

USA

Engine: Water-cooled, six-cylinder
Bore x stroke: 4.0 x 5.0in (100 x 125mm)
Capacity: 377ci (5,881cc)
PTO Power: 71.1hp @ 1,350rpm
Drawbar Power: 62.1hp @ 1,350rpm
Transmission: Six-speed
Speeds: 2.5-12.5mph (4.0-20.0km/h)
Fuel consumption: 8.8hp/hr per gallon
Weight: 8,625lb (3,881kg)

As related above, the late 1950s were catch-up time for Case – the company was busy producing its new four- and six-cylinder diesels, the all-new 400, 300 and 200. But amongst this flurry of activity, the biggest machine soldiered on. The 500 of 1953 was the first machine to use that new diesel engine, but otherwise is was little more than a revamped Model L, the basic design of which hailed back to 1929.

There were a couple of updates with the 600, which replaced the 500 in 1957 but was only produced in that year. Two extra speeds were added to the gearbox, giving six forward ratios in all and Flambeau Red gave way to a two-tone Desert Sunset/Red colour scheme. But really that was it. The same year, the short-lived 600 was itself replaced by the 900, which again was hardly changed apart from a taller, squared-off grille that incorporated the headlights. It became the 900B in 1958, to denote an injection pump change in the diesel, from a single-plunger American Bosch to a multi-plunger Robert Bosch.

There were of course variations on the 600/900 theme. The 620 and 920 were industrial models, with heavier front axles and softer seats. The tractor shown here is a 910B, the '10' revealing an LPG version of the six-cylinder diesel. Demand for LPG tractors was increasing though this one was outsold by the diesel equivalent by four to one.

Below: 910B was LPG-powered, but diesel would soon dominate.

Case 930 Diesel

USA

Engine: Water-cooled, six-cylinder
Bore x stroke: 4.125 x 5.00in (103 x 125mm)
Capacity: 401ci (6,257cc)
PTO power: 80.7hp @ 1,600rpm
Drawbar power: 70.9hp @ 1,600rpm
Transmission: Six-speed
Speeds: 2.5-13.6mph (4.0-34.0km/h)
Fuel consumption: 15.2hp/hr per gallon
Weight: 8,845lb (3,980kg)

Case had been very busy in the 1950s: the new diesel engine, Case-o-matic transmission (which added a torque converter to the existing 8-speed gearbox) and a whole new range of tractors. So by the early '60s it was heavily in debt, and settled in to a few years of consolidation and marketing, rather than engineering.

That's why the 30 series of 1960 brought little real change. Every tractor in the range acquired the new name, from 30hp 230 and 330, to 37hp 430 and 50hp 530, not to mention the 630, and 50hp 730/830. The 900, still based on the 1929 L model, became 930. But a couple of years later, the 930 did get a more substantial update, with the Comfort King. This isolated the driver's platform by moving it up, away from the chassis, and mounting it on rubber. As farmers didn't always put top priority on comfort, Case pushed the potential profitability of a smoother, more comfortable tractor – less fatigue, longer working hours and more profit. For whatever reason, the 930 Comfort King was a success, with or without a cab. So successessful, that this feature was extended to the 730 and 830 in 1964.

Below: Comfort King denoted Case's new rubber-mounted cab.

Case 1030

USA

Specifications for 1031 Diesel
Engine: Water-cooled six-cylinder
Bore x stroke: 4.375 x 5.00in (109 x 125mm)
Capacity: 451ci (7,036cc)
PTO power: 102hp @ 2,000rpm
Drawbar power: 88hp @ 2,000rpm
Transmission: Eight-speed
Speeds: 2.0-16.2mph (3.2-25.9km/h)
Fuel consumption: 13.24hp/hr per gallon
Weight: 9,335lb (4,200kg)

Mechanically, little had changed for the 930, still using Case's now well established 401ci six-cylinder diesel. But at 80-odd horsepower, it was in danger of getting left behind in the power race. Unllike the car and motorcycle market – undergoing a power race at the same time – this one had nothing to do with sheer speed. Not until the JCB Fastrac of the 1980s would high road speeds again become an issue in the tractor market. Instead, the search for

more tractor power was all about exacting a higher rate of work. If a tractor had, say, a 20 per cent power boost, that could turn it from a four-plough into a five-plough machine. Assuming the same speed, that meant 25 per cent more acres worked in a day. So the tractor power race wasn't simply a marketing man's tool, though there's no denying that it was a gift to tractor salesman, another way of putting one over on the opposition.

Case had already produced a response to this in the 930GP, putting together the 930's six-cylinder diesel with the 730 row-crop chassis. The next step was to turbocharge the engine – as Allis-Chalmers had pioneered in 1961 – or use the tried and tested method of making it bigger. So that's what Case did. The diesel's bore was increased by a quarter-inch to give a capacity of 451ci (7,036cc) – the five-inch stroke was unchanged. Combined as a new rated speed of 2,000rpm, that gave 102hp, the most powerful Case yet. Not to mention one of the first 100hp tractors –maybe marketing did have something to do with it after all.

Below: Most powerful Case yet – 102hp for the 7.0 litre 1030.

Case 870

USA

Specifications for 870 gasoline
Engine: Water-cooled, four-cylinder
Bore x stroke: 4.375 x 5.00in (109 x 125mm)
Capacity: 301ci (4,686cc)
PTO power: 71.1hp
Drawbar power: 62.5hp
Transmission: Eight-speed
Speeds: 1.9-15.0mph (3.0-36.0km/h)
Fuel consumption: 11.9hp/hr per gallon
Weight: 8,925lb (4,016kg)

In 1969, the same year that Tenneco increased its holding of Case shares to 99.1 per cent, the company announced an updated range of two-wheel-drive tractors. The new 70-series Agri-king line up would replace the 30-series Comfort King in its entirety. As it turned out, the smallest 470 and 570 were

dropped after a couple of years, as Case's acquisition of David Brown meant some rationalisation was essential. But for the bigger tractors, the 70-series would see Case through the 1970s.

All 70 series had a Comfort King type rubber-mounted platform to quell noise and vibration, and there was a choice of three different seats with adjustable suspension. Hydrostatic power steering was standard, as were rear-mounted non-metallic fuel tanks, which offered more capacity, better visibility and of course would not rust. The diesel engine had changed, now with more economical direct injection, though gasoline was still available on all except the big 1070. This was the last gasp for gasoline though, and LPG had already faded away – there was no LPG option on the 70 series. In the meantime, once the 470 and 570 had been dropped, the 770 gasoline was Case's smallest American-made tractor, with 53hp from its four-cylinder 251ci engine.

Below: 70-series was Case's new generation for the 1970s.

Case 970

USA

Specifications for 970 Powershift diesel
Engine: Water-cooled, six-cylinder
Bore x stroke: 4.125 x 5.0in (103 x 125mm)
Capacity: 401ci (6,256cc)
PTO power: 93.4hp @ 2,000rpm
Drawbar power: 79.9hp @ 2,000rpm
Transmission: Twelve-speed
Speeds: 1.8-17.0mph (2.9-27.2km/h)
Fuel consumption: 15.1hp/hr per gallon
Weight: 11,190lb (5,036kg)

This was the age of multi-speed transmission. At first, as the demand for more ratios increased, it was just a case of adding more gears to an existing conventional gearbox, which worked fine up to five- or six-speed. Then someone came up with the idea of a two-speed epicyclic gear set between

clutch and gearbox, doubling the ratios and enabling clutchless shifting. Most of the major tractor makers used this in some way, and gave it a fancy name – the Case version was the Case-O-Matic.

Powershift, which was an option right across the 70 series range, was an extension of the same priniciple. This used four mechanical gear ratios and three Powershift speeds, giving twelve forward speeds in all, and three reverse. The Powershift consisted of a planetary gear train and four wet disc clutches, which could shift on the move, though there was an interlock preventing forward-reverse selection unless the foot clutch was disengaged.

The 70 series also had self-adjusting disc brakes as standard, with power-assist as an option. A two-speed PTO – 540 and 1,000rpm – was available as well, thanks to a double-ended output shaft. To change speeds, you removed the shaft, turned it around and reinserted it. Finally, 1969 was the year that Old Abe retired. The faithful eagle had been the JI Case logo since 1865, but wasn't deemed sufficiently dynamic for the late 20th century.

Left: Twelve forward speeds and three reverse for the Case 970, plus a two-speed PTO. 'Agri King' was a generic name for the 70-series.

Case 1570 'Spirit of 76' USA

Engine: Water-cooled six-cylinder
Bore x stroke: 4.625 x 5.0in (116 x 125mm)
Capacity: 504ci (7,862cc)
Power: 180bhp @ 2,100rpm
Transmission: Twelve-speed

In 1976 – America's bicentennial year – red, white and blue patriotism gripped the nation. Even Harley-Davidson, that most American of symbols, acquiesced with suitably coloured bicentennial special editions of its heavyweight motorcycles. Tractor manufacturers were no exception, and Case produced a red, white and blue stars 'n stripes edition of the 1570 – the 'Spirit of '76'.

Appropriately for such a flag-waving tractor, this one was based on the flagship 1570, which actually debuted in 1976. It was actually the largest,

most powerful two-wheel-drive tractor of its time, with a turbocharged version of Case's 504ci (7,862cc) six-cylinder direct injection diesel. It produced 180bhp at 2,100rpm – that made it more powerful than the four-wheel-drive 1470 Traction King. A 12-speed transmission was standard, as was a cab with tinted windows. It also used the Agri King name, which was first used with Case's 70 series, and which went out of production in 1978.

The 70 series had its roots in 1969, as a new generation of Case tractors to emerge from several years of commercial uncertainty. They ranged from the small 470 (in both petrol and diesel form) up to the big 1070 for the 100-plus horsepower class – that led directly to the 1570. But the 1570 wasn't Case's only foray into special edition tractors – the 'Black Knight' was an elegant black, silver and red version of the 870.

Below: Turbocharger, 180bhp and a patriotic paint job for this 1570.

Case 2670 Diesel

USA

Engine: Water-cooled, six-cylinder, turbo intercooled
Bore x stroke: 4.625 x 5.0in (116 x 125mm)
Capacity: 504ci (7,862cc)
PTO power: 219hp @ 2,200rpm
Drawbar power: 189hp @ 2,200rpm
Transmission: Twelve-speed
Speeds: 2.0-14.5mph (3.2-23.2km/h)
Fuel consumption: 15.3hp/hr per gallon
Weight: 20,810lb (9,365kg)

However sophisticated a tractor's transmission, however big and chunky its tyres, there's a limit to how much power two wheels can transmit on a muddy, sticky field. The only answer was four-wheel-drive, and as the bigger machines approached and passed the 100hp mark, they often did so as 4x4s.

To meet this new breed of 100hp+ machines, Case announced the 1200 Traction King – someone in the Case marketing department was obviously a

royalist! This used a turbocharged version of the 451ci diesel first seen in the 1030. It produced 120hp, and powered a massive machine with four-wheel-drive, four-wheel steering and an all-up weight of 16,500lb. In the words of author P.W. Ertel (*The American Tractor*), 'The 1200 re-established Case as the pre-eminent builder of big wheatland tractors.'

In 1969, the 1200 was updated as a 70 series model, along with the rest of the range. Like the 1200, the new 1470 Traction King used hydrostatic steering for the front wheels and independent hydraulic power steering for the rear pair. This allowed the choice of crab steering, combined front and rear steering, and front or rear alone. It also had a disc brake operating on the main drive line. The biggest news though, was Case's new direct injection diesel, here in 504ci (7,862cc) form, with turbocharging. In 1972, Case's biggest tractor was replaced by the even more powerful 2470 with 176hp, and the 2670 with 221 PTO hp (256bhp gross) – both used the same six-cylinder diesel, now intercooled on the 2670.

Below: 2670 Traction King took Case into the four-wheel-drive market.

Case 2590 Powershift Diesel USA

Engine: Water-cooled, six-cylinder, turbo
Bore x stroke: 4.625 x 5.0in (116 x 125mm)
Capacity: 504ci (7,862cc)
PTO power: 180hp @ 2,100rpm
Drawbar power: 153hp @ 2,100rpm
Transmission: Twelve-speed
Speeds: 2.0-20.0mph (3.2-32.0km/h)
Fuel consumption: 13.3hp/hr per gallon
Weight: 15,740lb (7,083kg)

By 1978, Case had a far stronger European presence than ever before. The David Brown factories had been modernised and the old SFV tractor plant in France began building the 580F loader/backhoe. Case also took a 40 per cent stake in the French excavator and crane manufacturer Poclain, and began making Poclian equipment in the USA. Meanwhile, the 70 series was starting to look old, and Case responded with the 90.

Its biggest tractors had already started to use alternative engines – from the mid-'70s, the biggest-ever 2870 Traction King was powered by a 300bhp Scania engine. But the 'smaller' big tractors, such as the 2590 shown here, carried on with the company's home-built 504ci (7,862cc) six. They all had micro-electronic controls, which were now just starting to filter into tractor design. These started with the most expensive machines and filtered down, as systems got cheaper and market expectations rose. As well as the standard 4x4 90 series tractors, a specail 4690 was developed for Canadian conditions.
The 90 in turn was superseded by the 94-series in 1983, first as the two-wheel-drive range, covering 43-180bhp. The 4x4s soon followed, with the 4494 to 4994 range covering 210-400bhp (gross), the largest powered by a V8 turbo diesel. Hydra Shift was the latest transmission option now, which allowed full engine braking even as a semi-automatic. And there were electronics of course: an 'Intelligence Centre' monitored time, acreage worked and a wheel position indicator.

Left: 90-series replaced the Case 70 in 1978, introducing electronic control and the semi-automatic Hydrashift transmission.

Case 7130

USA

Specifications for Magnum 7130
Engine: Water-cooled, six-cylinder turbo-intercooled
Capacity: 532ci (8,300cc)
Power: 264hp
Transmission: 24-speed (+ 6 reverse)

Taking over International Harvesters in 1984 not only catapulted Case into second place in the US industry, it also presented it with a huge, overlapping range of tractors. International, more than Case, had sought to broaden its range by buying in machines from other manufacturers – everything from little Mitsubishi 4x4s to the gargantuan Steigers. Selling someone else's tractor invariably means less profit, and International had overstretched itself. Case had no intention of going the same way, and most of International's 95hp+ range was axed.

Specifically, the International 70-series four-wheel-drives, affectionately known as 'Super Snoopys' because of their overhanging nose, were dropped in favour of Case's 94 series and, in 1988, the Magnum 7000 series pictured here. The Magnum was and is a four-wheel-drive machine, but positioned half a class down from the Steiger-sized super-tractors. By 1997, five different models were on offer, all powered by Case's 532ci (8,300cc) six-cylinder turbo diesel, giving 155hp to an intercooled 264hp. All had four-wheel-drive, with kingpin steering on the smaller front wheels, Powershift, shuttle and a 24x6 (that's 24 forward, six reverse) transmission. The hitch control was electronic, as were many other features of the late '90s Magnum.

Below: The Magnum is Case's biggest conventional tractor, with engine power up to 235hp.

Case 7140

USA

Specifications for MX270 Magnum (2002)
Engine: Water-cooled, six-cylinder, turbo-intercooled
Capacity: 505ci (7,878cc)
Power: 235hp @ 2,000rpm
Torque: 1,802lb ft @ 1,400rpm
Transmission: 18-speed (+ 4 reverse)
Weight: 20,200lb (9,090kg)

The Case 7140 pictured here is an early Magnum, using that same Case-manufactured six-cylinder diesel. But it's interesting to compare the current range of 2002 Magnums with that of 1997 – the latest machines are no more powerful than those (less, in some cases) indicating that the sheer horsepower race is finally drawing to a close. Instead, advances in transmissions, electronics and engine efficiency are the chief weapons in building a competitive 21st century tractor.

So while the '97 Magnums offered 155-264hp, the current ranges from 145hp to 235hp. Four models were listed in 2002, the MX180, 200, 220 and

270. The top-range MX270 uses a four-valves per cylinder 7.9 litre six, with turbocharging and intercooling. Against the '97 Magnum's 24 forward speeds, it offers a 'mere' 18, but here again, electronic control is more than making up the difference.

But Magnums still occupy second-biggest spot in Case-International's five-strong line up of tractor ranges. These start with the DX Compacts, offering 21-39hp power units and with a wide range of specific implements; the CX Utility tractors occupy the 40-83hp range; MX Maxxums are 67-145hp tractors. And above the Magnums come the STX Steiger range. Case bought up Steiger in 1986, but 16 years later this name lives on, simply because it is so well known and respected among users of these super-tractors. All are four-wheel-drive of course, and the power range now covers 275 to 450hp. They're also available with Quadtrac, four caterpillar tracks in place of the double or triple wheel options sometimes used on machines of this size.

Below: Early Magnum – the latest ones use electronics as well as horsepower.

Caterpillar

USA

'She crawls along like a caterpillar' went an early slogan, and the name stuck. Today, Caterpillar is so strongly associated with tracked crawlers that its name is almost generic, in the way that 'Hoover' once was for vacuum cleaners. As far as farmers are concerned, crawlers are heavier and less manoeuvrable than wheeled tractors, yet they can keep going in heavy and wet soil where a conventional machine would bog down.

Daniel Best and Benjamin Holt both knew this, and both were pioneers in the crawler field. But they were bitter rivals, and even went to court over a patent dispute. However, the tough 1920s made clear that neither could survive on their own, and the two merged in 1925 to form Caterpillar. With new strength, the new company dominated the crawler market, moving

into construction and road making markets as well as agriculture. It pioneered the use of diesel in tractors, notably in the small D2, a crawler designed for farmers.

After World War II, Caterpillar consolidated its position as the world's leading maker of crawler machines, exporting them all over the globe. That was underlined by a joint venture with Mitsubishi in 1963. Caterpillar survived the early 1980s recession, but only at the cost of many jobs. However, it fought back in 1987 with the Challenger, an all-new rubber-tracked farm tractor. A new style of tractor had been born.

Below: Caterpillar – one of those instantly recognisable global brands.

Caterpillar Best Sixty

USA

Specifications for Best Sixty (1921)
Engine: Water-cooled, four-cylinder
Bore x stroke: 6.5 x 8.5in (163 x 213mm)
Capacity: 1,128ci (17,591cc)
PTO power: 56hp @ 650rpm
Drawbar power: 11,000lb (4,950kg)
Transmission: Two-speed
Weight: 17,500lb (7,875kg)

The Best company came out of World War I better than its rival Holt. The latter had concentrated on redesigning its crawlers to suit miltary and government use, which won plenty of orders during the war but left the company high and dry when peace came in 1918 – military specifications weren't necessarily suited to the civilian market. Worse still, when the war ended, the US Army crated up all the Holt tractors it had in France, shipped them home and sold them off at rock bottom prices.

By contrast Best had found big military contracts difficult to win. There was a good reason for this – the man they had to sell to at the War Department happened to be the cousin of Holt's vice president! They were also hampered by steel shortages after America entered the war in 1917. But while Holt was busy churning out tractors for the Army, Best was able to concentrate on the civilian market. So in 1918, it had well thought of machines, the 75, the Muley and new Model 40, all ready to sell. And it was working on another new crawler, the Sixty pictured here. This built on all the

Caterpillar 2 Ton

USA

Engine: Water-cooled, four-cylinder
Bore x stroke: 4.0 x 5.5in (100 x 138mm)
Capacity: 276ci (4,311cc)
Brake power: 25.4hp @ 1,000rpm
Drawbar power: 15.1hp @ 1,000rpm
Transmission: Three-speed
Speeds: 2.2-5.2mph (3.5-8.3km/h)
Fuel consumption: 9.45hp/hr per gallon
Weight: 4,040lb (1,818kg)

Caterpillar's 2 Ton started life as the Holt T-35 in 1921, and stayed in production after the 1925 merger with Best. It was effectively Holt's contribution to the first Caterpillar line-up. It was a small crawler, aimed at farmers rather than construction contractors.

Unusually, its four-cylinder gasoline engine used an overhead camshaft, a feature normally found on sports cars of the time. The 276ci (4,311cc) motor produced 15 drawbar hp, and 25 at the belt – disappointing for Holt, as the T-35 name derived from its expected horsepower. Another unusual feature was that the transmission (three forward speeds, one reverse) was located behind the rear axle, instead of in front. It used oil-cooled steering clutches, and the track links were cast.

Holt made almost every component of the 2 Ton, apart from the Eisemann magneto and KIngston carburettor, a practice continued by Caterpillar. One that did change though, was the price. An early advert for the T-35 (as it then was) quoted a price of just $375: '...a big and increasing demand requires increased

Above: Best Sixty was in production for 14 years.

lessons Best had learned in the previous ten years, and was something of a classic in crawler development. As a Caterpillar, it would stay in production until 1932.

Above: Brand new, just $375. The 2 Ton was aimed at farmers.

production and makes possible this reduction in price of the Supreme Small Tractor.' By 1927, the 2 Ton's penultimate year, that had risen to $1,850.

Caterpillar Thirty Distillate

USA

Engine: Water-cooled, four-cylinder
Bore x stroke: 4.5 x 5.5in (118 x 138mm)
Capacity: 350ci (5,456cc)
Brake power: 34.1hp @ 1,400rpm
Drawbar power: 26.7hp @ 1,400rpm
Transmission: Five-speed
Speeds: 1.7-5.4mph (2.7-8.6km/h)
Fuel consumption: 7.8hp/hr per gallon
Weight: 9,975lb (4,489kg)

This crawler, the Caterpillar Thirty, was highly successful, in production for over a decade and selling in large numbers. It was designed and produced by Best, launched in 1921 as the Model S, replacing the Model B.

There was nothing revolutionary about the Thirty, but its four-cylinder gasoline engine was strong and reliable. Originally, the four cylinders were cast separately, but later they were cast in pairs, with a common cylinder-head for each pair. These are known as 'two head' Thirtys. In either form, it was an overhead valve engine of 461ci (7,192cc), rated at 800rpm (later increased to 850rpm). When first tested by the University of Nebraska, it didn't actually make the 30hp suggested by its name (just like the T35 Holt, in fact) and a larger carburettor had to be fitted to make 18 drawbar hp and a genuine 30 at the belt. Within a few years, more engine modifications (higher speed, different cam timing and carburettor changes) produced nearly 38hp. Honour was satisfied.

Caterpillar Sixty

USA

Specifications for Sixty (1924)
Engine: Water-cooled, four-cylinder
Bore x stroke: 6.5 x 8.5in (163 x 213mm)
Capacity: 1,128ci (17,591cc)
PTO power: 72.5hp @ 650rpm
Drawbar power: Hp not measured – 12,360lb
Transmission: Three-speed
Speeds: 1.9-3.6mph (3.0-5.8km/h)
Fuel consumption: n/a
Weight: 20,000lb (9,000kg)

If the 2 Ton was Holt's contribution to the new Caterpillar company in 1925, then this, the big Sixty and mid-sized Thirty, was that of Best. When the two firms merged, their product line-ups naturally had to be trimmed to suit. Despite the long-standing rivalry, it all worked out rather well – Holt's 2 Ton was the better smaller crawler, while the Best Model S (Thirty) and Model A (Sixty) were superior to the ageing Holt 5-Ton and 10-Ton.

In fact, although lighter and with less power than the Holt Model 75 it replaced, the Caterpillar Sixty could actually do more work. It was a good design, agile for its size and notably well balanced. Like the smaller Thirty, it stayed in production for many years, in this case from 1919 to 1931. Power came from a large, slow-turning four-cylinder engine. It was rated at just 650rpm, but then each piston measured six and a half inches across! As standard, it ran on gasoline, but a kerosene option was available. Again like the Thirty, it failed to make its promised power when first tested by Nebraska,

Above: The Thirty was a popular crawler in fields and orchards.

Although the Thirty came without a canopy, one was available for farming use, and in fact there were special low-seat versions for orchard use – a power take-off was another option. In all its forms, the Thirty was a popular machine, and nearly 24,000 were built between Best and Caterpillar.

Above: Massive four-cylinder Sixty was Caterpillar's biggest of the '20s.

coming out at 35-55hp. But engine improvements brought 60hp, then 73hp, and a three-speed transmission replaced the original two-speeder. Nearly 14,000 Sixtys were built by Best and Caterpillar.

Caterpillar Sixty-Five

USA

Engine: Water-cooled, four-cylinder
Bore x stroke: 7.0 x 8.5in (175 x 213mm
Capacity: 1,308ci (20,401cc)
Brake power: 72hp @ 650rpm
Drawbar power: 54hp @ 650rpm
Transmission: Three-speed
Speeds: 1.9-4.4mph (3.0-7.0km/h)
Fuel consumption: 8.2hp/hr per gallon
Weight: 24,965lb (11,234kg)

In 1929, there were 48 tractor manufacturers in the USA. Four years later, fewer than ten were left. The depth and speed of the depression following the 1929 Wall St Crash shocked everyone, and tractor makers were hit hard – sales slumped to the lowest figures since 1915. Farms went bust, construction projects were cancelled – in 1932, Caterpillar posted a loss of $1.6 million.

But it survived, and within a few years of that million-dollar loss was not only making profits, but could claim to be the world's largest producer of diesel engines. There were several reasons why Caterpillar survived: the Soviet Union was a valuable export market; despite increased competition from wheeled tractor makers, Caterpillar was still the crawler leader; and of course, it was a pioneer of diesel power. A big crawler could use up to 100 gallons of gasoline a day, but a modern diesel would use half that. Little wonder that the Sixty-Five shown here was Caterpillar's last new gasoline tractor – and it lasted only three years, from 1932-35.

But the Diesel Sixty which ran alongside it, rapidly updated as the Diesel Sixty-Five, was a real success. Within two years of its arrival in 1931, Caterpillar was making more diesels than all other US tractor makers combined.

Below: Sixty-Five was Caterpillar's last gasoline crawler.

Caterpillar R2

USA

Engine: Water-cooled, four-cylinder petrol
Bore x stroke: 4.0 x 5.0in (100 x 125mm)
Capacity: 251ci (3,916cc)
Brake power: 29.2hp @ 1,250rpm
Drawbar power: 21.9hp @ 1,250rpm
Transmission: Three-speed
Speeds: 2.0-3.6mph (3.2-5.8km/h)
Fuel consumption: 8.9hp/hr per gallon
Weight: 7,420lb (3,339kg)

The Caterpillar shown here, the R2, is a rarity – only eighty-three were built over three years. It was small by crawler standards, a 6,000lb machine which underlined again how responsive Caterpillar was to the demands of the market.

It was part of a new generation of new and updated smaller crawlers, which Caterpillar came up with to meet increased competition from tractor

manufacturers. So the Fifteen was an update of the earlier Ten, and was built in 1932 and 1933. A bigger Twenty (itself a re-rated version of the 'big' Fifteen) ran alongside it.

Meanwhile, the old Flathead Twenty was treated to a new overhead valve engine in 1934. With a bore and stroke of 4.0in x 5.0in (100 x 125mm), it had a capacity of 251ci (3,916cc). Powered thus, it was renamed the Twenty-Two and stayed in production right up to 1939, a mark of its popularity. The R2 pictured here was just a different version of the same thing. In the mid-1930s, government orders arrived for Caterpillars to work in President Roosevelt's New Deal construction projects. Built to this specification, they were renamed R2 (was Twenty-Two), R3 (Twenty-Eight) and R5 (Thirty-Five and Forty). The R stood for Roosevelt. The R2 only lasted until 1937, but Caterpillar used the name again for the gasoline equivalent to the new D2 diesel.

Below: Caterpillar R2 was part of President Roosevelt's New Deal.

Caterpillar D2

USA

Specifications for D2 Diesel (1939)
Engine: Water-cooled, four-cylinder
Bore x stroke: 3.75 x 5.0in (94 x 125mm)
Capacity: 221ci (3,444cc)
Brake power: 30hp @ 1,525rpm
Drawbar power: 25.1hp @ 1,525rpm
Transmission: Five speed
Speeds: 1.7-5.1mph (2.7-8.2km/h)
Fuel consumption: 13.2hp/hr per gallon
Weight: 7,420lb (3,339kg)

While diesel power was proving such a hit with the loggers and road makers, Caterpillar hadn't forgotten that farmers could benefit from it as well. So with one eye on established tractor makers, it announced the D2 in 1939 – diesel power as standard, but the right size for the average farmer. 'The Diesel D2 gives you Power, Traction, Economy.'

It came in two track widths – 40 inches and 50 inches (1-1.25m) – and the four-cylinder diesel engine measured 252ci (3,931cc), enough for 42hp at 1,150rpm. A two-cylinder donkey engine was used to start up, though 24-volt electric starting was optional. There were standard and low-seat orchard versions.

But the D2 wasn't startlingly new, more a consolidation of Caterpillar's diesel dominance. Throughout the 1930s, a string of new diesels had been running off the production lines at the Preoria factory. The Diesel Fifty of 1933 was the first, with a 5.25 x 8.0in (131 x 200mm) four-cylinder engine, and a four-speed transmission. The Diesel Thirty-Five (1933-34 only) used a three-cylinder version of the same motor, though its roots went back to the original Best/Caterpillar Thirty. That was replaced by a Diesel Forty, which included major updates to the chassis. All of these had gasoline equivalents, but not for much longer – Caterpillar would not be making any gasoline crawlers after 1945.

Below: D2 marked the arrival of a diesel option for farmers.

Caterpillar Twenty-Eight

USA

Specifications for Twenty-Five (1932)
Engine: Water-cooled, four-cylinder
Bore x stroke: 4.0 x 5.5in (100 x 138mm)
Capacity: 276ci (4,306cc)
Brake power: 33hp
Drawbar power: 6,011lb (2,705kg)
Transmission: Three-speed
Speeds: 1.8-3.6mph (2.9-5.8km/h)
Fuel consumption: 9.84hp/hr per gallon
Weight: 8,087lb (3,639kg)

In the tough economic conditions of the early 1930s, Caterpillar came under increasing pressure from fresh competition. Faced with falling sales of conventional tractors, and keen to diversify, many of the wheeled tractor manufacturers were looking to move into the crawler market. Allis-Chalmers

had already done so with its Monarch, and Cletrac was well established. International Harvester, too, was considering crawlers. For all of them, it made sense, as crawlers could use the same basic design as a wheeled tractor – a new product for minimal cost.

This was the reason behind Caterpillar's string of updated and new crawlers in the early 1930s. Despite tough trading conditions, it had no choice but to spend money on modernising its product line. So the little Ten was re-rated as a Fifteen, and the old Fifteen became known as the 'Flathead Twenty' – both thanks to a power boost.

Meanwhile, the old overhead valve Twenty had been uprated into the Twenty-Five in 1931. Two years later, it was updated again, with engine improvements bringing 28hp at the drawbar. This was the Caterpillar Twenty-Eight pictured here. As a government-ordered tractor, it became the R3.

Below: Caterpillar had to update fast to meet new rivals in the '30s.

Caterpillar RD6

USA

Specifications for RD-7 Diesel
Engine: Water-cooled, four-cylinder
Bore x stroke: 5.75 x 8.0in (144 x 200mm)
Capacity: 831ci (12,964cc)
Brake power: 68hp @ 850rpm
Drawbar power: 61hp @ 850rpm
Transmission: Four-speed
Speeds: 1.6-4.7mph (2.6-7.5km/h)
Fuel consumption: 14.9hp/hr per gallon
Weight: 21,020lb (9,459kg)

Caterpillar product names, the reader can hardly have failed to notice, can be a little confusing. Or at least, they were in the 1930s, when a plethora of new models added to the mix. The 'Twenty' and 'Flathead Twenty' for example, are completely different machines. As well as the new models, the company seemed torn between the tried and tested method of

referring to models by their horsepower (Fifteen, Sixty and so on) and the new simpler nomenclature brought in by the government-specification machines – the R2, R3 and R5.

In 1936, Caterpillar finally did away with the old system altogether. From then onwards, all tractors would use the R prefix, or RD if they were diesels. So the big Diesel Fifty became the RD7; the new small R4 was joined by a diesel RD4, which had 35hp at the drawbar and used a twin opposed cylinder engine for starting. The RD6 pictured here was simply a new name for the existing Diesel Forty, which had been launched two years previously. This in turn was an update of the Diesel Thirty-Five, powered by the same three-cylinder engine, but finally removing the last links with the old Caterpillar Thirty. There was also a bigger RD8, with 118 belt hp and over 20,000lb of pull available at the drawbar. The RD6 managed 56hp and a pull of just under 9,700lb.

Below: The RD6 was simply a renamed Diesel Forty.

Caterpillar D4 Diesel

USA

Engine: Water-cooled, four-cylinder
Bore x stroke: 4.5 x 5.5in (113 x 138mm)
Capacity: 350ci (5,456cc)
PTO power: 47hp @ 1,400rpm
Drawbar power: 33hp @ 1,400rpm
Transmission: Five-speed
Speeds: 1.7-5.4mph (2.7-8.6km/h)
Fuel consumption: 11.4hp/hr per gallon
Weight: 11,175lb (5,029kg)

The Caterpillar D4 arrived in 1938, but really it was no more than the RD4 with a single top roller – the RD4 used two. The RD4 had been launched two years earlier, as the diesel version of the new Caterpillar Thirty, with a two-cylinder donkey engine to start it.

The D4 was used by the US Army during World War II, with armour plating attached. But this was only a small part of Caterpillar's war work. Much

production was allocated to the Lend-Lease agreement, under which useful equipment was shipped over to England. Once the US entered the war in December 1941, some models were discontinued to make room for production of tank transmissions and cannon shells. But the military also ordered the D7 and D8 crawlers in quantity – the D7 was particularly in demand as it woudl fit into a landing craft. In fact, Caterpillar couldn't keep up with the orders, and the government contracted American Car & Foundry to produce 1,000 of them under licence. One interesting project was converting the Wright Cyclone aero engine to take diesel fuel – it had been developed as a tank engine, but was too thirsty. Caterpillar did this successfully, but the result was heavier and bulkier than the original – the Army rejected it after just seventy-five had been delivered.
As for the D4, that resurfaced after the war in 59hp form. In 1955, tests at Nebraska showed it to have 9,976lb (4,489kg) of drawbar pull and a fuel consumption of 13.94hp/hr per gallon.

Below: The D4 saw war service, complete with armour plating!

Caterpillar Challenger

USA

Specifications for Challenger 65 (1987)
Engine: Water-cooled, six-cylinder
Power: 270hp
Transmission: 10-speed Powershift
Weight: 31,000lb (13,950kg)

In the 1930s, '40s and '50s, Caterpillar was successful in persuading farmers of the advantages of crawlers over wheeled tractors. But times changed, and by the mid-1990s,four-wheel-drive tractors had largely taken over field work from the crawlers. If that weren't enough, double- or triple-rear wheel options offered far greater traction and flotation. Both had far greater road speed than any crawler. What was needed was a cross between the two.

In 1986, it arrived. The all-new Caterpillar Challenger used high-flotation rubber tracks in place of the traditional steel, giving reduced soil compaction,

and faster quieter travel on tarmac roads. The company had been experimenting with the idea for several years. Rubber track conversions were tried on the existing D6, D3 and D4, and tested on all sorts of terrain alongside conventional four-wheel-drive wheeled tractors. The tests were evidently successful, as the Challenger went into production in 1987. Fifteen years later, it's still with us.

In fact, that first Challenger made an interesting comparison with a comparable articulated four-wheel-drive tractor of the same power (270hp). It weighed nearly 30 per cent more and cost $140,000 – $30,000 more than its rival. But there were enough customers out there willing to pay the extra. In early 2002, manufacturing rights to the Challenger were sold lock, stock and caterpillar tracks to AGCO.

Below: The rubber-tracked Challenger pioneered a new form of crawler.

Caterpillar Challenger 85D USA

Specifications for 35 (2002)
Engine: Water-cooled, six-cylinder, turbo diesel
Power: 149hp
Track Width: 18in (450mm)
Weight: 26,750lb (12,038kg)

Variations on the Challenger theme soon followed, though it was five years before a specific row-crop versions appeared. In the meantime, the 65B, C and D were based on the 270hp original, but the Challenger 75 of 1990 had a more powerful 325hp turbo diesel.

More significant for farmers though were the Challenger 210hp 35 and 240hp 45, unveiled in 1994. Smaller and lower powered than the first machines, these had higher ground clearance, and the tread width could be adjusted between 60 and 88 inches. The tracks themselves were available in 16- or 32-inch widths. Both machines had an electronically controlled 16-speed

Powershift transmission that allowed sequential shifting one gear at a time, automatic sequential shifting, shifting to a pre-selected gear, programmed shifting, or use of the upper six speeds only. A 265hp 55 joined the range in 1996.

The 85D pictured here was also launched that year, and has another twist in sophisticated transmissions. To reduce torque stresses on the final drive, and reduce slippage, power was limited tp 360hp in gears three and four, and to 355hp in first and second. Only in gears five to ten could you have the full 370hp. But even the 85D wasn't the most powerful Challenger – a 410hp 95 joined the range in 1999. By then, Challengers were already appearing in the green and white colours of Claas of Germany. Under a marketing agreement, Claas agreed to market and service Challengers in Europe, under its own name.

Below: Challenger 85D had a variable power limiter to save the transmission.

Claas Jaguar 690

GERMANY

Specifications for Jaguar 900
Power: 605hp
Cutter head width: 29.5" (738mm)
Knives on drum: 24
Feed and compressor rolls: 4
Optional: Mechanical four-wheel-drive

Claas is a German company, a well-established maker of combine harvesters. It started developing combines, intended for European conditions, in 1930, which it does to this day – all-round machines as well as specialists for crops such as sugar cane. Altogether, it claims that its range of combines can cope with over 80 different crops.

The Jaguar 690 pictured here is one of the large forager rangers, aimed at whole crop silage. A wide range of such combines ranges from the 321hp 830 to the 600hp+ 900. On the 900, auto steering and lubrication, and a metal detector are standard. Alongside these, Claas makes a large range of balers: the Quadrant turns out large square bales, the Rollant does round bales and the Variant, as its name suggests, is a variable chamber baler. It was Claas that came up with net wrapping for round bales in the mid 1980s. It also produces a range of pulled and mounted foragers from its Saulgau plant. Claas has never made tractors, but its name is used on Caterpillar machines.

Below: Not a tractor maker, but Claas has been building combines for 80 years.

Cletrac

USA

The Cleveland Motor Plow Company (hence Cletrac) was a pioneer of tracked tractors. It was formed in 1916 by Rollin H. and Clarence G. White. From the start, they intended to produce tracked tractors for the farm trade, which was something of an innovation. Another innovation – which remained unique to Cletrac – was differential steering. The conventional means of steering a tracked vehicle was to declutch the inside track, which means losing traction.

Cletrac fitted a plantary gearset controlled by a brake at each drive cog, which allowed sending controlled amounts of power to each track continuously. Cletrac was bought out in 1945 by the Oliver Corporation, which carried on using the name until 1965.

Below: It wasn't just Caterpillar – Cletrac was the crawler pioneer.

Cletrac K15-25

USA

Engine: Water-cooled, four-cylinder
Bore x stroke: 4 x 5.5in (100 x 138mm)
Capacity: 276ci (4,306cc)
Brake power: 28.4hp @ 1,375rpm
Drawbar power: 4,375lb (hp not measured)
Transmission: Three-speed
Speeds: Not tested
Fuel consumption: 9.2hp/hr per gallon
Weight: 4,775lb (2,170kg)

The first Cletracs were advertised as being 'Geared to the ground' to emphasise their superior traction over conventional wheeled machines. Running continuous tracks over cogged wheels had been pioneered in World War I by early military tanks, but in the 1920s they rapidly gained in popularity for agricultural duties. Cletrac, with one exception, never made anything else.

One of its earliest offerings was the Model W 12-20, of which 17,000 were built between 1919 and 1932. A Model F followed in 1920, in four versions, including the small Hi-Drive 9-16, designed for small farms. The F was driven by a floating roller chain between drive gears and tracks. The K-20 shown here was built between 1925 and 1932, and unlike earlier Cletracs the track was lubricated with a manual plunger. It used Cletrac's own four-cylinder engine.

Right: The K15-25 was an early, small Cletrac.

Cletrac 60-80

USA

Engine: Water-cooled, six-cylinder
Bore x stroke: 5.5 x 6.5in (138 x 163mm)
Capacity: 617ci (9,625cc)
Brake power: 90hp @ 1,050rpm
Drawbar power: 83.5hp @ 1,050rpm
Transmission: Three-speed
Speeds: 1.8-3.6mph (2.9-5.8km/h)
Fuel consumption: 12.1hp/hr per gallon
Weight: 22,840lb (10,382kg)

Most early Cletracs were relatively small – a bigger Model 30A (rated at 30-45hp) went on sale in 1926, but was built only in limited numbers. But 1933 saw the debut of the 60-80 pictured here. It was Cletrac's first diesel-powered machine, using a six-cylinder Hercules engine, and a departure from the kerosense/gasoline motors used until then. Like all Cletrac diesels, it had electric start. The controlled differential steering using band brakes allowed both tracks to be driven while turning, which improved traction in sticky terrain. In 1945, Cletrac was bought by Oliver, though production continued in White's old Cletrac factory and the tractors themselves were little changed, albeit with the Oliver badge. But history came full circle in 1960, when Oliver itself was bought up by White, which once again found itself in charge of Cletrac. Production was moved to the old Hart-Parr works in Charles City, Iowa, but ceased altogether in 1965.

Right: The big 60-80 was Cletrac's first diesel.

Cletrac

80
Cletrac
Cletrac

Cockshutt

CANADA

Cockshutt was that rare thing, a Canadian tractor manufacturer. Established in Ontario in 1877 by J. G. Cockshutt, it began selling Hart-Parr tractors in the early 1920s, but didn't begin making its own until 1945. But this turned out to be the company's swansong. The White Motor Corporation bought Cockshutt in 1962, and enforced close co-operation with Oliver. There were no more 100 per cent

Cockshutt tractors from then on, and after about ten years of badge-engineering the Cockshutt name faded out altogether. A shame, as the genuine Cockshutts were beautiful looking machines.

Below: Shortlived – Cockshutt was independent for less than two decades.

Cockshutt 40

CANADA

Specifications for Cockshutt 40 petrol
Engine: Water-cooled, four-cylinder
Bore x stroke: 3.4 x 4.1in (85 x 103mm)
Capacity: 230ci (3,588cc)
PTO power: 38.7hp @ 1,650rpm
Drawbar power: 30.3hp
Transmission: Six-speed
Speeds: 1.6-12.0mph (2.6-19.2km/h)
Fuel consumption: 9.7hp/hr per gallon
Weight: 5,305lb (2,411kg)

Cockshutt never made its own engines, being too small to make this economically viable. Instead, it bought in units from Buda, Perkins and Hercules. This wasn't unique among tractor manufacturers, though even the Buda range was off-limits to Cockshutt after the engine manufacturer was taken over by Allis-Chalmers in the '50s.

Announced in 1949, the 40 used a six-cylinder Buda, and was Canada's biggest tractor at the time. This engine came powered by gasoline, distillate or diesel, and no more nor less than the four-cylinder Buda with two extra cylinders tacked on. Rated at 1,650rpm, it achived nearly 39hp in Nebraskan tests. The gearbox was six-speed, with two reverse, and hydraulics and a live PTO were optional. There was a Cockshutt 50 as well, essentially exactly the same machine, but with a slightly larger version of the same Buda six. All of them – 20, 30, 40 and 50 – were also sold under the Co-Op brand – Cockshutt

Cockshutt 20

CANADA

Specifications for Cockshutt 20 petrol (1952)
Engine: Water-cooled, four-cylinder
Bore x stroke: 3.2 x 4.4in (80 x 110mm)
Capacity: 140ci (2,184cc)
PTO power: 27.4hp @ 1,800rpm
Drawbar power: 20.2hp @ 1,800rpm
Torque: 186lb ft
Transmission: Four-speed
Speeds: 2.5-13.3mph (4.0-21.2km/h)
Fuel consumption: 10.5hp/hr per gallon
Weight: 2,813lb (1,279kg)

'One of the cutest and best performing little tractors ever made,' according to author Robert Pripps. The 153ci Cockshutt 30 was the Canadian company's first home-built tractor, and was joined after a few years by the smaller 20.

The 20 was designed as a natural rival for that other neat small tractor, the Allis-Chalmers B. So it used a 124ci (1,934cc) Continental engine (Cockshutt never built its own engines) which was soon dropped in favour of a 140ci (2,184cc) – both were four-cylinder units with 4.4in (110mm) strokes, and both came in gasoline or distillate form. Compression ratios were quite high for the day (5.0:1 distillate, 6.75:1 gas) and at the rated 1,800rpm the bigger 20 produced just over 27 PTO hp at tests in Nebraska. Hydraulics and a live PTO were optional, and standard transmission was four-speed. The Cockshutt 20 was also sold as the Co-op E2.

Above: Six-cylinder 40 was Cockshutt's biggest tractor.

had taken over the National Farm Equipment Co-Op in 1954. Cockshutt later hired the famed industrial designer Raymond Lowey, and he duly came up with the squared off 500 Series. But was it really any better looking than the neat, home penned originals?

Above: 'One of the cutest and best?' Cockshutt's little 20.

Co-op

CANADA

Engine: Water-cooled, six-cylinder
Bore x stroke: 3.4 x 4.1in (85 x 103mm)
Capacity: 230ci (3,588cc)
PTO power: 38.7hp @ 1,650rpm
Drawbar power: 30.3hp
Transmission: Six-speed
Speeds: 1.6-12.0mph (2.6-19.2km/h)
Fuel consumption: 9.7hp/hr per gallon
Weight: 5,305lb (2,411kg)

Co-op tractors were no more nor less than Canadian Cockshutt machines, repainted and rebadged to suit. There was nothing new in this. The Allis-Chalmers U model of 1929 – one of the most successful A-Cs ever – was

originally built at the request of a Chicago-based farmers' co-operative. When the deal fell through, Allis built the new machine for itself.

In fact, Cockshutt's first in-house tractor,the 30 of 1945, was marketed by both Canada's farm co-operative as the CCIL 30 and by the American Farmers Union Co-op as the Co-op E3. The latter was treated to a repaint of Pumpkin Orange. Mechanically, it was identical to the Cockshutt.

The arrangement must have worked, for that was followed by the Co-op E4, based on the Cockshutt 40 but again with the corporate orange livery. Cockshutt's little 20hp 20 was sold as the Co-op E2 and the big 50 as the E5. The latter was powered by a 273ci (4.3 litres) six-cylinder engine from Buda, in gasoline or diesel forms.

Below: The Co-op was sold by both Canadian and US co-operatives.

David Brown Cropmaster UK

Specifications for 1949 diesel
Engine: Water-cooled, four-cylinder
Bore x stroke: 3.5 x 4.0in (88 x 100mm)
Capacity: 154ci (2,402cc)
Power: 34hp @ 1,800rpm
Transmission: Six-speed

David Brown became a tractor maker almost by accident. His company, based in Huddersfield, England, was a gear specialist, until Harry Ferguson asked him to produce the transmission for his new tractor with it's revolutionary three-point hitch. Brown ended up building the complete Ferguson tractor, from the mid-1930s until the two men parted company a couple of years later.

Evidently taken with the tractor business, David Brown began producing his own machine, the VAK-1, from 1939. It had a hydraulic lift similar to the

Ferguson, but was a bigger, more powerful tractor. Many were built during World War II for military use. But what really made the David Brown name in tractor circles was the Cropmaster, launched in 1947.

This was a comprehensive update of the VAK-1, notable for including many features as standard, which at the time cost extra (or weren't available at all) on most tractors. The hydraulic lift, a swinging drawbar and eletric lights were all standard. Coil ignition, a two-speed PTO and six-speed gearbox also featured. One interesting point – unique on tractors of the time – was the wind deflector, a modest attempt to protect the driver from the elements! Also forward looking was the diesel option from 1949. This 34hp direction-injection unit produced 23hp at 1,800rpm and featured wet cylinder liners. The Cropmaster was replaced in 1953.

Below: Note the wind deflector on this Cropmaster, a classic David Brown.

David Brown 990

UK

Specifications for 990 Diesel
Engine: Water-cooled, four-cylinder
Bore x stroke: 3.6 x 4.5in (90 x 113mm)
Capacity: 186ci (2,902cc)
PTO power: 51.6hp @ 2,200rpm
Drawbar power: 44.4hp @ 2,200rpm
Transmission: Six-speed
Speeds: 2.1-14.1mph (3.4-22.6km/h)
Fuel consumption: 15.2hp/hr per gallon
Weight: 4,770lb (2,147kg)

In some ways, the Cropmaster was ahead of its time, and in fact the David Brown company was capable of looking beyond the bounds of conventional tractors. The smaller 2D of 1955 was anything but conventional, with its air-cooled twin-cylinder engine mounted behind the driver, and the tool carrier powered by compressed air.

A range of more conventional machines followed through the 1950s and '60s, such as the modern looking 900 in 1956, which offered a choice of four-cylinder diesel or gasoline engines of 42.5hp. The 990 Implematic shown here was introduced in 1961, with the diesel option now up to 52hp. This was the first David Brown with a cross-flow cylinder-head and two-stage front-mounted air cleaner. It was updated again in 1963, with height control for the Implematic hydraulics, a longer wheelbase and a new 12-speed transmission option. The smaller 850 and three-cylinder 770 were added to the range as well. The 55hp 990 Selectamatic was unveiled in 1965, and carried through to 1971 – a four-wheel-drive option was available from 1970.

David Brown was taken over by Tenneco, owners of Case, in 1972, and all DB machines wore Case badges from 1983. Five years later, Case-International closed the factory down.

Below: David Brown later made more conventional machines.

Deutz MTZ320

GERMANY

Specifications for Deutz MTZ222 (1927)
Engine: Water-cooled, single-cylinder
Power: 14hp
Weight: c6,000lb (2,700kg)

Deutz was a pioneer of diesel engines, indeed of internal combustion engines in general, but not of tractors. Co-founder Nicolaus Otto was experimenting with four-stroke engines in the 1860s, and his company produced its first diesel in 1898, only a few years after Rudolph Diesel completed his first prototype. Deutz made military tractors during WWI, and an agricultural machine in 1919, and soon afterwards brought the two together – the first Deutz diesel tractor entered production in 1926.

It was a semi-diesel in the familiar European mould, the sort of machine

built for years by Lanz, Landini, Hurliman and Marshall. The MTZ320 pictured here was simply a later development of the same thing. It had been preceded by the MTZ220 in 1931, which used a twin-cylinder engine of 30hp. They worked on the usual semi-diesel principle, which was perhaps the simplest form of internal combustion engine available. European semi-diesels such as the Deutz MTZ were certainly simpler than the typical American tractor of the same period. But unlike later Deutz tractors, these had water-cooled engines, notably the F1M 414, introduced in 1936 and which finally ceased production in 1951 after 19,000 had been built. It was the last water-cooled Deutz tractor – from then on, until the 1990s, all would use air-cooled diesel engines.

Below: All Deutz tractors were diesel powered, first as semi-diesels.

Deutz D8006

GERMANY

Specifications for 8006 Diesel
Engine: Air-cooled, six-cylinder
Bore x stroke: 3.94 x 4.72ci (99 x 118mm)
Capacity: 354ci (5,522cc)
PTO Power: 85.5hp @ 2,100rpm
Drawbar Power: 69.5hp @ 2,100rpm
Transmission: Sixeen-speed
Speeds: 0.6-17.2mph (1.0-27.5km/h)
Fuel consumption: 16.9hp/hr per gallon
Weight: 6,850lb (3,0833kg)

Deutz finally entered the US tractor market in 1966, alongside its range of diesel engines, which were used in a variety of applications. All of these were air-cooled, which made them unique in the '60s, though the tractors were quite well received. The tractors tended to find more favour in colder northern climates, where the inability of the air-cooled engine to freeze up was much appreciated. In the south though, there were cases of overheating.

Apart from being air-cooled, the new-to-America Deutz tractors were relatively conventional, with 8- or 16-speed transmissions. Among the 1971 line-up were the D5506 (with a 49hp 3.3 litre/212ci four-cylinder engine) and the D8006 pictured here. The latter was a 73hp tractor, using a 5.5 litre/354ci six-cylinder motor, and was popular on dairy farms. The example shown here was still at work on a farm in Wisconsin in the mid-1990s, a quarter-century after rolling off the Deutz production line. It was a quality product.

One other thing set Deutz tractors apart from the opposition. Being air-cooled, they were noiser, and when the University of Nebraska tested a D8006 in 1971, it scored 98dB(A) on the noise meter, compared to 91dB(A) for the equivalent International. With noise (both inside and outside the cab) becoming a hot issue, this was something Deutz would have to address.

Right: Air-cooled Deutz tractors such as this D8006 were noisy, but wouldn't freeze up or boil over.

DEUTZ
D 80 06

Deutz 6260

GERMANY

Engine: Air-cooled, four-cylinder diesel
Capacity: 263ci (4.1 litres)
Power: 71hp
Transmission: 12-speed

When Deutz took over the Allis-Chalmers tractor business in 1985, it had good reasons for doing so – access to the Allis dealer network, and that long-established brand name. For the American market, Deutz tractors were renamed Deutz-Allis (though the same machines were still badged Deutz-Fahr for Europe). The colour remained Deutz green.

Typical of that late '80s range was the 6200 series, which came in vineyard, orchard, low profile or standard formats. The 6200 had the choice of two- or four-wheel-drive and a 65hp 3.8 litre/244ci or 71hp 4.1 litre/263ci engine – both

of course were Deutz air-cooled diesels. Like other low/mid-size tractors of the time, it had hydrostatic steering and a four-speed gearbox with three ranges, giving twelve forward speeds in all. Air conditioning was standard on the flat floor Star Cab.

But in the last decade of the 20th century, noise and emissions regulations began to tighten for tractors. Deutz could see that its air-cooled engines would have difficulty meeting the new standards, and its answer came in 1996, two months after the company was taken over by SAME, in fact. The futuristic Deutz Agrotron replaced 80 per cent of the entire range. With its low sloping bonnet and huge glassy cab, this looked the part of a 21st century tractor. The engine choices were 68-95hp four-cylinder units, and sixes up to 145hp – all of them water-cooled.

Below: Into the late 1980s, Deutz persevered with air-cooling.

Deutz-Allis 9150

GERMANY/USA

Specifications for 5.125
Engine: Air-cooled, six-cylinder
Capacity: 393ci (6,128cc)
Power: 120hp @ 2,300rpm
Torque: 41kg m @ 1,600rpm

The late 20th century was a turbulent time for the tractor industry. An agricultural slump in the USA, and difficult trading conditions just about everywhere, led to closures, mergers, take-overs and management buy-outs. The tractors badged Deutz-Allis were just one product of these difficult times. As a name, Deutz-Allis didn't last long, but it formed the basis of the giant AGCO company, that now includes a whole list of famous names.

Deutz was German of course, a pioneer of both internal combustion in

general and diesels in particular – its first diesel tractor was announced in 1926. It became a leading maker of diesel tractors, and took over Fahr, another German make, in 1968 to become Deutz-Fahr. The company developed a range of air-cooled diesels, which were almost unique in the tractor market – the Porsche tractor also used an air-cooled engine. By the mid-1980s, Deutz was keen to find a way to expand its share of the North American tractor market. Meanwhile, the ailing Allis-Chalmers was looking to sell its tractor business – so in late 1985, Deutz-Allis was born.

The tractors, such as the 9150 shown here, were sometimes a hybrid of Deutz air-cooled diesel engines and American-made components.

Below: Hybrid. The first Deutz-Allis tractors used both US and German components.

Deutz-Allis 9130

GERMANY/USA

Specifications for 5.100
Engine: Air-cooled, five-cylinder
Capacity: 327ci (5,107cc)
Power: 95hp @ 2,300rpm
Torque: 33kg m @ 1,600rpm

Although the American market was in recession, the Allis takeover made sense for Deutz, with access to the big A-C dealer network, not to mention a well-respected name. Its first act was to close down the old Allis factory just before Christmas 1985. Then it badged its own tractors as Deutz-Allis before restarting production in the USA four years later.

As mentioned above, these were hybrid tractors, using a combination of Deutz
air-cooled diesel engines in American chassis – the 9130 pictured here was just

such a hybrid, with front-wheel-assist and a twin-wheel capability.

It also used a Deutz air-cooled engine, which remained a Deutz speciality. Although this system promised lower maintenance and a quicker warm-up, the air-cooled motors could overheat in very dusty conditions. But the real problem was noise – Deutz-Allis tractors were far noisier than their water-cooled rivals, and this was rapidly becoming a big issue in tractor design. In any case, the Deutz-Allis brand was short-lived. In 1990, the American manufacturing operation was transformed into AGCO by a management buy-out. AGCO grew rapidly, buying up White-New Idea in 1991 and Massey-Ferguson three years later. The Deutz-Allis machines were renamed AGCO-Allis, but the Deutz influence lingered – those air-cooled diesels weren't phased out until 1996.

Below: 9130 was available with front-wheel-assist, a cheaper form of four-wheel-drive.

Doe

UK

Specifications for Triple D (1959)
Engine: Water-cooled, four-cylinder (x2)
Bore x stroke: 3.94 x 4.52in (99 x 113mm)
Capacity: 440ci (6,864cc)
Transmission: six-speed
Weight: c. 5 tons (4,455kg)

Today, 100hp four-wheel-drive tractors are nothing special, but in mid-1950s England, they didn't exist. So farmer George Pryor built one himself, by joining two Ford Majors together, one behind the other. Result – an instant four-wheel-drive machine with a combined power of 104hp. In 1958, it was put into production by local Ford dealer Ernest Doe, who for years had been producing specialist Ford conversions. The famous Doe Triple D was born.

Compared to contemporaries, the Triple D had superb traction and performance, thanks to the extra power and near perfect weight distribution.

And, despite its ungainly appearance, it was astonishingly manoeuvrable – it measured over 20ft long, but thanks to a near-90-degree articulation had a turning circle of 21ft, some 5ft less than the standard Ford Major. At £1,950, the Triple D was also far cheaper than a big crawler (a 93hp Caterpillar cost £7,000). Of course, inexperienced drivers could come to grief – it was possible to select a forward gear on one tractor and reverse on the other! But the Triple D was a success, with 289 sold and exported all over the world, including the USA. It was replaced in 1965 by the 130hp Doe 130, based on two Ford 5000 tractors.

But within a few years the Doe tandem tractor was no more. Mainstream manufacturers were now delivering more power, not to mention four-wheel-drive, and they could do it more profitably than a small concern like Doe. Instead, this family firm went back to selling tractors, as Colin Doe (grandson of Ernest) does to this day.

Below: Triple D offered 100hp and four-wheel-drive in the 1950s.

Eagle 20-35

USA

No specifications available

'Eagle – The Tractor that Takes the Guesswork Out of Farming!' Never undersold, the Eagle was a pioneer machine – the Eagle Manufacturing Company had been building farm equipment for many decades, but jumped into the new tractor market in 1906 with a twin-cylinder machine. Several models were offered over the early years: a 12-22, 13-25 and 16-30, plus the big 20-35 pictured here.

'You take no chances when you buy an Eagle tractor...tested and tried out in every condition of field and belt work...it will do your work at a big saving over

horse cost, and therefore will pay you a profit...Built right, works right and is priced right...You get superior engine qualities – the famous Eagle two-cylinder engine with three-bearing crankshaft, cylinders cast in pairs, valve-in-head action which gives more power and cleaner combustion...Perfex radiator insures thoro cooling even in hottest weather. Hyatt high duty roller bearings throughout transmission. Double drive insures perfect traction under all conditions...Simple in construction...any man can run it and keep it in repair without help...Write today for illustrated folder and full details.'

Below: Big 20-35 was Eagle's largest tractor.

Eagle 6C

USA

Specifications for 6A Gasoline
Engine: Water-cooled, six-cylinder
Bore x stroke: 4 x 4.5in (100 x 113mm)
Capacity: 226ci (3,527cc)
PTO power: 40.4hp @ 1,416rpm
Drawbar power: 27.9hp @ 1,416rpm
Transmission: Three-speed
Speeds: 2.5-4.5mph (4.0-7.2km/h)
Fuel consumption: 9.2hp/hr per gallon
Weight: 5,670lb (2,577kg)

Eagle wasn't just wedded to the twin-cylinder concept, and had introduced a four-cylinder tractor in 1911 – a whole range of fours were offered through the

1920s. But it was an increasingly outdated line-up, especially compared to the Fordson. Next to Henry's mass-produced baby, the Eagles were heavy, expensive and old-fashioned.

In 1930, it finally responded with the relatively modern 6A, powered by a six-cylinder Waukesha engine producing 40 belt hp and nearly 28 at the drawbar. There were three of them. The 6A pictured here was designed for three- or four-plough use, the 6B was a row-crop machine and the 6C a utility. These might have been a big step for Eagle, but they weren't enough for the company to resume production after World War II. Tractor production ceased in the early 1940s, reportedly because engine supplies dried up, thanks to the war effort.

Below: Eagle bought-in a Waukesha engine for its six-cylinder 6C.

Emerson-Brantingham 12-20 USA

Engine: Water-cooled, four-cylinder petrol
Bore x stroke: 4.75 x 5.0in (119 x 125mm)
Capacity: 354ci (5,522cc)
PTO power: 27hp @ 900rpm
Drawbar power: 17.6hp @ 900rpm
Transmission: Two-speed
Speeds: 2.1-2.8mph (3.7-4.5km/h)
Fuel consumption: Not measured
Weight: 4,400lb (2,000kg)

Emerson-Brantingham was a pioneer of the American agricultural machinery industry, though it didn't take that name until 1909. John H. Manny was an inventor who produced a successful horsedrawn reaper in 1852. A couple of years later, financier Ralph Emerson joined the company followed by Charles Brantingham. In 1912, Emerson took over the Big Four Tractor Company of

Minneapolis. Big Four's Model 30 was a typically huge machine for the time, with 30hp at the drawbar – it was uprated to 45hp after that. Such big machines were losing favour, one reason why Big Four came up for sale relatively cheaply.

But Emerson also made smaller machines. The Model 20, introduced the year after Big Four came under the Emerson wing, was one such. That was followed by the 12-20 in 1918, a thoroughly conventional four-cylinder machine with two-speed transmission and a weight of 4,400lb (2,000kg). It was also far more powerful than the official rating of 12-20 might suggest – at Nebraska, the tractor tested mustered over 17hp at the drawbar and 27 at the PTO! The 12-20 was updated as the Model K in 1925. The Model Q, pictured here, came later. But financially Emerson-Brantingham never really recovered from the 1920s farming depression – JI Case bought the company in 1928, and soon dropped its tractor line.

Below: 12-20 was a rare machine – made more power than E-B claimed!

Ferguson

UK

Harry Ferguson and Henry Ford were joint fathers of the modern tractor. While Ford mass-produced tractors cheap enough that nearly any farmer could afford, Ferguson invented the three-point hydraulic hitch, the biggest milestone in tractor history. This took the manual drudgery out of coupling and decoupling implements; it also made use of the implement's drag as downforce on the tractor's rear wheels, improving traction; there was draft control too; and it prevented the tractor flipping over backwards, a serious safety problem in the early days. Seventy years on, every tractor on the market has a three-point hitch.

But, though Ferguson was a near-genius, he was an inventor, not a manufacturer, and had to enter into agreements with established tractor makers to get his invention on the market. Nothing wrong with that except that

the four agreements he made, one after the other – with David Brown, Ford, the Standard Motor Co and Massey-Harris – all ended in disagreements, and in the case of Ford, a three-year multi-million dollar law suit.

Ferguson was probably a difficult man to work with, but his legacy to the modern tractor is huge. Not only did his three-point hitch revolutionise the use of implements, but he had the sheer determination to see it into production, whatever problems arose. When he died in 1960, Harry Ferguson was negotiating with Massey-Ferguson to produce a new Ferguson tractor, with flat-four engine and torque converter transmission.

Below: Harry Ferguson revolutionised tractor design.

Ferguson-Brown

UK

Specifications for Ferguson Model A
Engine: Water-cooled, four-cylinder
Bore x stroke: 3.25 x 4.0in (81 x 100mm)
Capacity: 133ci (2,075cc)
Power: 20bhp @ 1,400rpm
Transmission: Three-speed
Speeds: 1.6-4.9mph (2.6-7.8km/h)
Wheelbase: 69in (1,725mm)

Harry Ferguson didn't just invent the three-point hitch, but a whole series of tractors to go with it. He perfected the three-point hitch in the early 1930s, and designed a whole tractor to suit it. The Ferguson Black (so-called because of its Henry Ford-esque colour scheme) looked the part of a revolutionary tractor, with its wide stance, low seat and streamlined fuel tank. It was powered by an American Hercules side-valve engine, a simple four-cylinder unit of 133ci (2,075cc). The ignition was US-made too, by American Bosch magneto.

Now Ferguson needed someone to build it. English engineering firm David Brown agreed to do it, and the Ferguson Model A started rolling off the production lines in 1936. It was little different to the Black, though used a Coventry Climax four-cylinder engine in place of the Hercules – this was reportedly a licensed version of the Hercules anyway, which David Brown took over producing itself, after about 250 Model As had been built. There was a three-speed gearbox, the choice of steel wheels or pneumatic tyres, and adjustable wheel treads.

Ferguson T20

UK

Specifications for Ferguson TE20
Engine: Water-cooled, four-cylinder
Bore x stroke: 3.19 x 3.75in (80 x 94mm)
Capacity: 120ci (1,872cc)
PTO power: 22.59hp @ 2,000rpm
Drawbar power: 16.35hp @ 2,000rpm
Transmission: Four-speed
Speeds: 2.9-11.5mph (4.6-18.4km/h)
Fuel consumption: 8.99hp/hr per gallon
Weight: 2,760lb (1,242kg)

Ferguson's next partner was Henry Ford, and the famous 'handshake agreement' between these two farmers' sons took place in 1939. Ford would mass produce a new tractor – the 9N – complete with Ferguson hitch, while Harry would market it. In some ways, it was a success – over 300,000 N tractors were built over the next eight years, but Ford reputedly lost money on every single one. Henry Ford II lost no time in ending the Ferguson agreement.

That had only covered North America, and in the meantime Harry had found someone who would build a Ferguson-equipped tractor for him in Britain. In 1945, the Standard Motor Company had a large factory in Coventry standing idle – it had produced aircraft engines during the war. An agreement soon followed, and as with David Brown and Henry Ford, Standard agreed to build the tractor, leaving Ferguson to market and sell it.

They worked fast, and only a year later the first Ferguson TE20 rolled off the production line at Banner Lane – the factory, incidentally, is used by

Above: Fergie-Brown was built by David Brown Engineering.

But after only two years, distrust had grown up between David Brown and Harry Ferguson. The manufacturing agreement ended, and Harry started searching for a new partner.

Above: The 'little grey Fergie' was Britain's answer to the Fordson.

Massey-Ferguson to this day. It was the 'little grey Fergie' familiar to a couple of generations of European farmers, and now regarded with the same sort of affection as the original Fordson, Farmall or John Deere are in America. In looks, the TE20 closely resembled the Ford 9N, though it had the useful update of a four-speed gearbox, and used an overhead valve Continental engine.

Ferguson TEA20

UK

Engine: Water-cooled, four-cylinder
Bore x stroke: 3.2 x 3.7in (80 x 92mm) (later 3.4 x 3.7in/85 x 92mm)
Capacity: 113ci (127ci) (1,841cc/2,070cc)
PTO power: 23.9hp (28.2hp)
Transmission: Four-speed
Speeds: 2.5-9.75mph (4.0-15.6km/h)
Weight: 2,376lb (1,069kg)

In the late 1940s, Britain was impoverished, deeply in debt after a six-year war. It needed to import as little as possible and export everything it could. So it was always in the gameplan that the TE20's American-made Continental engine would be replaced by Standard's own, when that was ready.

Unusually, it was a car engine, albeit a simple, torquey four-cylinder unit

eminently suitable for tractor use. It had been designed for the Standard Vanguard car, specifically intended for export. After 48,000 Ferguson TE20s had been built, the Standard-engined TEA20 was phased in during 1948. It was famously launched at the Claridges Hotel in London, driven down the steps by Harry Ferguson himself!

These were difficult years for Ferguson – the legal battle with Ford was ongoing, and he reacted by setting up his own tractor factory in Detroit, to produce machines for the American market. This wasn't just a challenge to Ford – by now Ferguson was responsible for a large distribution company in North America, which had no tractors to sell. The TO-20 was an Americanised version of the TE20, and was followed by powered-up TO-30 and 35.

Below: Unusually, the TEA20 was powered by a car engine.

Ferguson TED20

UK

Engine: Water-cooled, four-cylinder
Bore x stroke: 85 x 92mm (3.4 x 3.7in)
Capacity: 127ci (1,981cc)
PTO power: 23.9hp (later 25.4hp)
Transmission: Four-speed
Speeds: 2.5-9.75mph (4.0-15.6km/h)
Weight: 2,376lb (1,069kg)

Simple as the little grey Fergie looks, it was available in a whole range of different formats. As well as the basic field tractor, it could be had in vineyard form, with a narrow tread and as a full semi- or basic industrial tractor. Then there were the various fuels on offer: TEA signified gasoline, TED vapourising oil, TEH lamp oil and TEF diesel. The TED shown here has a proprietory four-

wheel-drive conversion, though of course all Fergusons of this period left the factory with two-wheel-drive.

Vapourising oil (TVO) was a sort of blended paraffin (kerosene) and much cheaper than gasoline in Britain because it wasn't taxed. Harry Ferguson reportedly resisted making a TVO-spec tractor, but he was persuaded that to do otherwise would be economic suicide in the cost-conscious UK tractor market. TVO (like the distillate available in the USA) produced less power than gasoline, so to compensate, the Standard engine's bore was increased by 5mm (0.2in) to 85mm (3.4in), which actually put it in line with the Standard Vanguard car – the gasoline TEA20 later received this larger engine as well. The diesel TEF joined the range in 1950, using a diesellised version of the Standard power unit.

Below: Four-wheel-drive conversion on this TED20.

Ferguson 35

UK

Specifications for 35 diesel (1957)
Engine: Water-cooled, four-cylinder
Bore x stroke: 3.4 x 4.1in (84 x 102mm)
Capacity: 45ci (2,259cc)
Power: 37bhp @ 2,000rpm
Transmission: Six-speed
Speeds: 0.96-10.5mph (1.5-16.8km/h)

This, the Ferguson 35, was a comprehensive update of the little grey Fergie – although planned and designed before the merger with Massey-Harris in 1953, it didn't go into produciton until three years afterwards. It had actually been preceded by the Detroit-built TO-35, which enjoyed the same updates.

Specifically, there was more power (still with the choice of gasoline, TVO or diesel), improved hydraulics with position control, and a six-speed gearbox. It

was also restyled, though for some reason the Detroit machines retained the TE20 styling. Another update came in 1959, after the takeover of diesel engine manufacturer Perkins – the Standard diesel was replaced with the three-cylinder Perkins, the 3.152, giving 35hp. It was slightly less powerful than the Standard, but easier to start. The Ferguson 35 came in grey and gold colours and was updated again in 1962 as the 35X. Now it had 'Multi-power' which effectively doubled the number of gear ratios available, and a more powerful version of the Perkins triple.

And in 1998, you could still buy a brand new Fergie... Well, almost. The Smallholder Tractor Co refurbished old TE20 or MF35 skid units, fitted new Lister-Petter diesel engines and sold the resulting mix of old, reconditioned and new parts as a T20 or ST35. Could this be the Morgan of tractors?

Below: Better hydraulics and more power for the Fergie 35.

Ferguson 40

UK

Engine: Water-cooled, four-cylinder
Bore x stroke: 3.19 x 3.88in (80 x 97mm)
Capacity: 134ci (2,090cc)
PTO power: 31.1hp @ 2,000rpm
Drawbar power: 24.4hp @ 2,000rpm
Transmission: Six-speed
Speeds: 1.2-14.6mph (1.9-23.4km/h)
Fuel consumption: 11.34hp/hr per gallon
Weight: 3,432lb (1,544kg)

This, the Ferguson 40, was the last tractor to bear the Ferguson name on its own. It was also one of the first fruits of the new Massey-Harris-Ferguson combine. Keen to make the most of the two tractor ranges, the new company painted the little Massey-Harris Pony grey and sold it through Ferguson dealers. Meanwhile, the Detroit-built Ferguson TO-35 was stretched, repainted and

relaunched as the Massey-Harris 50. This left Ferguson dealers unhappy, for they had nothing bigger to sell than the 35. To placate them, the Ferguson 40 was unveiled, basically an M-H 50 repainted in cream and grey.

In theory, this gave both M-H and Ferguson dealers a full range of tractors to sell without the expense of designing new machines. In practice, it simply confused the buying farmers and was in danger of damaging hard-won brand loyalty. Within a few years, there was a properly unified range of Massey-Fergusons.

One casualty was the planned Ferguson TE60, which Harry Ferguson had on the drawing board. This was a 48hp machine with a 190ci (2,964cc) gasoline or 224ci (3,494cc) diesel engine, designed to fit in the range above the TE and TO series. When the TE60 development didn't go ahead after the merger, it was the last straw for Harry – he sold his shareholding back to Massey-Harris, and resigned.

Below: The Ferguson 40 was also sold as a Massey-Harris 50.

Fiat 315

ITALY

Specifications for Fiat 702
Engine: Water-cooled, four-cylinder
Capacity: 400ci (6,235cc)
Power: 30hp (on gasoline)
Transmission: Three-speed
Weight: 5,720lb (2,600kg)

Ask any tractor enthusiasts to reel off the names of half a dozen manufacturers, and Fiat is unlikely to be among them. But this Italian company is one of the largest tractor makers in the world. How it achieved that was simple: the financial backing of the huge Fiat empire at home in Italy; a determination to export; and the acquisition of the Ford-New Holland concern in 1991, which gave access to the North American market.

Fiat's first tractor wasn't a real pioneer machine, though the company was

experimenting with tractors as early as 1910. World War I got in the way, while Fiat was diverted into military work, but a prototype was finally unveiled in October 1918. It went into production the following year, as the 702, the first Italian tractor to go into full production. The 702 was a mixture of existing parts: it had unit construction (with the engine and transmission as load-bearing members, as pioneered by Wallis and Fordson). The engine though, came straight out of a Fiat 3½ ton truck. This 6.2 litre four-cylinder unit was tough and well proven, giving 25hp on kerosene, 30hp on gasoline.

The post-war 315 shown here was several generations of Fiat tractor down the line, but it would never have happened without the first 702.

Below: 315 was a late 1960s small tractor. Home-made screen is a later addition.

Fiat 680DT

ITALY

Specifications for Fiat 580DT diesel
Engine: Water-cooled, three-cylinder
Bore x stroke: 4.06 x 4.33in (102 x 108mm)
Capacity: 168ci (2,621cc)
PTO power: 51.5hp @ 2,700rpm
Drawbar power: 42.1hp
Transmission: 16-speed
Speeds: 0.3mph-15.9mph (0.5-25.4km/h)
Fuel consumption: 13.91hp/hr per gallon
Weight: 6,630lb (2,984kg)

The 680DT shown here was a typically modern Fiat tractor of the 1970s, launched in 1977. It was part of a whole range of turbo-diesel tractors, and specifications are givén for the slightly smaller 580DT, which was tested at Nebraska in 1978. That range culminated in the four-wheel-drive 1000DT, Fiat's

first 100hp wheeled tractor, and the 150hp 1300DT. At the time, Fiat claimed to be the largest manufacturer of 4x4 tractors in the world with a broad span of 45-150hp offered. It was also about to extend that further by marketing Canadian Versatiles in Europe, with Fiat badges and colours.

Such a future must have seemed very distant to Fiat in 1945, when it sold only 32 agricultural tractors to a war-ravaged Italy. The company's saviour came in the form of the tiny 18hp 600 of 1949. Just as baby cars like Topolino and later 500 proved to be Fiat's backbone, so this baby tractor got Fiat back into mass production. Sales rocketed to 4,000 that year, and by 1959 were over 20,000, many of these four-wheel-drive machines. Fiat was also heavily involved in crawler manufacture, having produced what it claimed was Europe's first mass-produced crawler in 1932. Growth through the '60s and '70s was fuelled by expansion into foreign markets: by the '70s, Fiat tractors were being made in Turkey, Romania, Spain, Argentina and what used to be Yugoslavia.

Left: Fiat sold just 32 tractors in 1945 – now it's one of the biggest manufacturers in the world.

Fiat 88-94

ITALY

Engine: Water-cooled, four-cylinder turbo-diesel
Capacity: 250ci (3.9 litres)
Power: 85hp @ 2,500rpm

By 1980, Fiat was the quintessential multi-national player. As well as having its machines licence-built all over the world, it had a good working relationship with Allis-Chalmers. Fiat's 50hp tractor had been sold in the USA with Allis badges since 1976, and the company had jointly built earth-movers with the Americans since 1974. The marketing arrangement with Versatile had just kicked off, and by acquiring smaller manufacturers (Braud, Laverda and Toselli) it now had a full line-up of machines. Of the 100,000-odd tractors it built a year in the 1980s, 90 per cent were exported – Fiat lay claim to 13 per cent of the entire world market. Fiat was a true multi-national – not in its ownership, but its operations.

Given this global outlook, it was hardly surprising when Fiat took over Ford

New Holland in 1991. There were good reasons for doing it, not least of which was that the Allis-Chalmers arrangement was over, since that firm had been taken over by Deutz. This latest acquisition ensured Fiat's access to the massive US tractor market, through an established dealer network and respected brand name. It also offered the chance to extend the Fiat range with large modern machines. So if Fiat's G series of the 1990s looked like a Ford 70 series with new badges and paintwork, that's because it was.

But Fiat's own Italian-bred tractors carried on as well. The little 66 series offered three- and four-cylinder machines from 35hp upwards, while the 88-94 pictured here was Fiat's mid-range tractor, an updated version of the 93 series which ran alongside it.

Below: New Holland badges shows this 88-94 to be a post-takeover model.

Ford

USA/UK

The Ford tractor story (as opposed to the Fordson) started in 1939, with the famous handshake agreement between Henry Ford and Harry Ferguson. This brought the three-point hitch to market, and was a true landmark in tractor history. With its Ferguson hitch as a major selling point, the 9N and later 2N sold in huge numbers.

But Ford lost a lot of money on all of them – Henry's grandson Henry II (now head of the giant family firm) tore up the agreement with Ferguson on the way to making Ford tractors profitable again. Something else changed too – old Henry's vision had been to offer a single model of tractor that just about everyone could afford. But the market was changing, and of necessity Ford

introduced bigger, more sophisticated and complex tractors, built on both sides of the Atlantic. The UK and US sides were integrated in the early 1960s, joined by a range of mini-tractors imported from Japan ten years later. Ford-badged Steigers, meanwhile, gave Ford an entree to the super-tractor market. New Holland was taken over in 1985, and Versatile two years later, and by the early '90s there were very strong links with Fiat. In fact, the Italians were to buy the whole caboodle, linking up with Case in 1999. But the Ford legacy – the mechanisation of thousands of farmers – lived on.

Below: Ford tractors took up where Fordson took off.

Ford 9N

USA/UK

Specifications for Ford 9N (1939)
Engine: Water-cooled, four-cylinder
Capacity: 199ci (3,104cc)
PTO power: 20hp
Drawbar power: 12.7hp
Transmission: Three-speed
Weight: 2,340lb (1,063kg)

With the Fordson F in 1918, Henry Ford revolutionised the tractor market. Twenty-one years later, he did it all over again with the 9N. Of course, he did have a little bit of help the second time round. Irishman Harry Ferguson had much in common with Ford: another farmer's son, a self-made man who could be difficult and irascible, but was also close to genius. All these traits mirrored those of Henry Ford.

Ferguson had designed a revolutionary new means of attaching tractors to

implements. His three-point hydraulic hitch (fully patented) enabled quick and easy coupling and decoupling; its clever geometry allowed some of the implement's drag to be applied as downforce to the tractor's rear wheels, improving traction; it had draft control; and it was far safer than a conventional hitch, as it prevented the tractor rearing up and flipping over backwards – in the days before ROPS cabs, that could be fatal.

He demonstrated the system to Henry Ford, who was so impressed that he immediately agreed to build an all-new tractor incorporating the three-point system. The 9N of 1939 was the result, and it had much the same impact as the original Fordson F had done. Like the F, it was small, light and possessed an excellent power to weight ratio – it could do the work of a tractor of twice the price and weight. Ford and Ferguson didn't coin the phrase 'size isn't everything', but they might as well have done.

Below: 9N was a combination of Ferguson ingenuity and Ford manufacturing.

Ford 2N

USA/UK

Specifications for Ford 2N (1941)
Engine: Water-cooled, four-cylinder
Capacity: 199ci (3,104cc)
PTO power: 20hp
Drawbar power: 12.7hp
Transmission: Three-speed
Weight: 2,340lb (1,063kg)

The 9N was an immediate success. Quite apart from the obvious advantages of Ferguson's ingenious three-point hitch, it was exceptionally easy to use, and unthreatening to first-time tractor drivers. At the launch of the 9N, eight-year-old David McLaren was lifted into the seat, and proceeded to plough perfect furrows! The little tractor had a low seat and low centre of gravity; to start it, you just pressed a button; and it was so quiet that Ford suggested a radio could be fitted.

But even as the first production 9Ns were rolling off the line in mid-1939, war in Europe was almost a reality. When it broke out, even North America was hit by a shortage of raw materials, which in turn hit production. For the duration of the war, Ford was forced to introduce a simplified version of its new tractor, the 2N. This replaced the rubber tyres with steel wheels (already an option on the 9N anyway) and sacrificed the battery and generator electrics in favour of a magneto – that meant a return to hand cranking, but hell, there's a war on.

Production slumped from over 40,000 tractors in 1941 to fewer than 16,500 the following year, when the lines actually stopped for several months. But as soon as the war was over, the 9N, complete with rubber tyres and electric start, was back

Ford 8N

USA/UK

Specifications for Ford 8N (1947)
Engine: Water-cooled, four-cylinder
Bore x stroke: 4.2 x 3.5in (105 x 88mm)
Capacity: 155ci (2,418cc)
PTO power: 38hp @ 2,000rpm
Drawbar power: 33hp @ 2,000rpm
Torque: 118lb ft @ 1,150rpm
Transmission: Ten-speed (optional)
Speeds: 1.0-16.4mph (1.6-26.2km/h)
Weight: 3,790lb (1,723kg)

But behind the success of the 9N, there was trouble. Henry Ford and Harry Ferguson had made their famous 'handshake agreement' in 1938 – no paperwork, no lawyers. That was fine when things were going well, but in 1945 Mr Ford was 82 years old and in poor health. He finally retired, allowing his 28-year-old grandson Henry II to take over the reins. The young Henry had some tough decisions to make. Ford the company was losing money, especially in tractors: the N series tractors had lost over 25 million dollars in six years.

Henry II lost no time in getting to the root of things. Central to the handshake agreement was that Ford sold all its tractors to Harry Ferguson, who marketed and distributed them. Now he was given nine months notice – from mid-1947, Ford would do that job itself. Ferguson was furious, and took Ford to court, eventually winning over $9 million in compensation.

But for Ford it was worth it to win control of distribution and thus profits. Part of the plan was an updated 9N – the new 8N (announced to coincide with

Above: The 2N was a wartime austerity version of the 9N.

in production. In fact, 1946 was its best ever year, with 74,004 tractors built.

Above: The 8N finally made a profit for Ford.

the new regime) had 20 design improvements over its predecessor, including a four-speed gearbox, better brakes and a position control for the hydraulic lift. It was a huge success, and over 100,000 were sold in the first full year, outselling its nearest rival by ten to one. Even better news for young Henry, each one of them made a profit for Ford.

Ford NAA/600 Series

USA/UK

Specifications for Ford NAA (1953)
Engine: Water-cooled, four-cylinder
Capacity: 134ci (2,090cc)
PTO power: 27.5hp

Ford's 600-series tractors, introduced in 1954, were a couple of generations down from the 8N, though related to it. The 8N, itself a direct descendant of the first Ferguson-Ford 9N, had been radically transformed into the NAA the year before. Also known as the Golden Jubilee (Ford was celebrating 50 years in the business), it looked superficially similar to the 8N, but was both heavier and more powerful. For the first time in a Ford or Fordson tractor, there was an overhead valve engine – more powerful and fuel-efficient than side-valves. There was new sheet metal too, though the 8N's grey and red colour scheme was retained. What the NAA

didn't have was high-clearance or a diesel option – that's why Ford imported the British-built Fordson into America alongside its new homegrown tractor.

The NAA lasted only a year in production, being replaced by the 600 and 800 series in 1954. The 600 (such as the 601 Workmaster pictured here) was powered by the same 134ci (2.1 litres) motor as the NAA. But there were transmission and ancilliary advances: the new 640 had a four-speed transmission; the 650 was a five-speed; and the 660 added a live PTO. And for those who wanted more power, the 800 offered the same line-up of options with a 172ci (2.7 litres) version of the same engine. A year later, row-crop versions of these two were offered, the 700 and 900 series, but there was no diesel option until 1959.

Below: In the '50s, Ford moved away from the single-model philosophy.

Ford 6000

USA/UK

Specifications for Ford 6000 petrol (1961)
Engine: Water-cooled, six-cylinder
Bore x stroke: 3.62 x 3.60in (91 x 90mm)
Capacity: 223ci (3,479cc)
PTO power: 66.7hp @ 2,400rpm
Drawbar power: 63.1hp
Transmission: Ten-speed
Speeds: 1.2mph-18.2mph (1.9-29km/h)
Fuel consumption: 10.32hp/hr per gallon
Weight: 7,225lb (3,284kg)

Nineteen sixty-one was a significant year for Ford tractors. After 30 years of separate operation, the UK and US ends of the operation were brought together. Until then, they had taken their own decisions. It was the British arm, for example, that declined to put the Ford-Ferguson 9N straight into production at Dagenham. The British factory built its own line of diesels,

which were shipped across the Atlantic, but there was no attempt to co-ordinate the UK and US ranges – even the colours were different, and blue/orange British tractors sat alongside their grey/red stablemates in Ford dealers. This all changed in 1961, with the 'World Tractor' programme. The two colours were combined to form one corporate identity, and a unified model range was developed.

The 6000 tractor, announced that year, was the first fruit of this change. It was a mid-range row-crop machine, using a 223ci (3.5 litres) six-cylinder gasoline engine or 242ci diesel. Unfortunately, it also used the troublesome Select-O-Matic transmission, which Ford had introduced a couple of years before. This allowed the driver to shift through all ten gears without stopping the tractor, but in the field it just wasn't durable enough. Despite various recalls and updates, it remained in production until 1967.

Below: Ford streamlined its US and UK ranges – this 6000 was American-made.

Ford 3000

USA/UK

Specifications for Ford 3000 diesel
Engine: Water-cooled, three-cylinder
Bore x stroke: 4.2 x 3.5in (105 x 88mm)
Capacity: 155ci (2,418cc)
PTO power: 38hp @ 2,000rpm
Drawbar power: 33hp @ 2,000rpm
Torque: 118lb ft @ 1,150rpm
Transmission: Ten-speed (optional)
Speeds: 1.0-16.4mph (1.6-26.2km/h)
Weight: 3,790lb (1,723kg)

Ford's small 3000 series filled a gap in the new unified range, giving a complete range of tractors. Moving into the '70s, the company also showed increasing interest in producing specialised tractors, such as golf course groomers, which until then had usually only been available from specialised firms, often converting a Ford base tractor.

The range was certainly more rationalised now. The same year that the new 6000 was announced, the existing Fordson Super Major was renamed 5000; the old 800/900 series was now the 4000 and the older 600/700 was now 2000, with the UK-built Dexta as the 2000 Diesel. Meanwhile, Ford USA's own small diesel was dropped. Four years later, the gap between 2000 and 4000 was filled by...you've guessed it. The new 3000 used a three-cylinder gasoline or diesel engine, and was given modern squared-off styling to match an updated 4000. The little 2000 was updated as well, with a 158ci (2.5 litres) petrol or diesel

Ford 7810

USA/UK

Specifications for Ford 7810 diesel (1978)
Engine: Water-cooled, six-cylinder
Capacity: 401ci (6,256cc)
PTO power: 86hp
Transmission: 16-speed (optional)

Ford's 7810 was introduced in 1978, but had roots going back ten years. It used a detuned version of the six-cylinder diesel first seen in the 8000 of 1968. This tractor was a wide-front row-crop machine with Ford's own engine producing 106hp and a choice of 8- or 16-speed transmissions. A turbocharged version with 131hp – the 9000 – soon followed. But in the 7810 it was detuned to produce 86hp, an interesting means of achieving the required output, and possibly more long-lived than turbocharging a smaller four-cylinder engine! A 16-speed transmission was among the options, as well as front-wheel-assist. The latter was a popular option at this time, offering some of the benefits of four-wheel-drive without the cost and complication of a full 4x4 system.

Two years previously, Ford had dropped the original 7000 series in favour of the 84hp 7600, which ran alongside the 70hp 6600. Counting down the 1976 range from there was the 60hp 5600, 52hp 4600, 45hp 4100, 40hp 3600 and 32hp 2600. If that wasn't small enough, Ford had also started importing mini-tractors from Japan, and labelling them as Fords. The 1600 was one such – it had first appeared as the 1000 in 1973, a twin-cylinder machine of 23hp weighing only 2,300lb.

***Above:* Three-cylinder 3000 slotted in below 4000, as gasoline or diesel.**

engine (three cylinders again) and the option of 6- or 12-speed transmissions. The 3000 was replaced in 1976 by the 40hp 3600, still in petrol or diesel form, and a variety of chassis types, though the 'all-purpose' was most popular.

***Above:* 7810 used a detuned six-cylinder diesel of 86bhp – 131bhp in 9000 turbo.**

Fordson

USA/UK

Henry Ford came from farming stock, and wanted to popularise tractors by mass-producing them, just as the Model T had done for cars. The Fordson (he couldn't use his own name on a tractor, as it was already in use at the time) did exactly that, and established Ford in the tractor market. But it was as much a European story as an American one. Fordson tractors were soon being built in Cork, Ireland, and Dagenham, England, with the updated Model N crossing the Atlantic to sell in the USA.

In fact, Henry Ford stopped making tractors in America in 1928 – he was clearing the decks to make room for the vital new Model A car, which was to replace the Model T. But production continued apace in England, where

the Fordson acquired quite a following. Fordsons had been a feature of English agriculture since World War I, when the British governent ordered 6,000 to help with wartime food production. They had a huge impact – it's estimated that before 1914 a mere 500 tractors were at work on Britain's farms. The legacy was a strong line of Fordson tractor manufacture and use in Britain, which in fact went on using the Fordson name right up until the early 1960s.

Below: Fordson mechanised thousands of farmers, by making a cheap tractor.

Fordson F (1917)

USA/UK

Specifications for Model F Kerosene (1920)
Engine: Water-cooled, four-cylinder
Bore x stroke: 4.0 x 5.0in (100 x 125mm)
Capacity: 251ci (3,916cc)
PTO power: 18.1hp @ 1,000rpm
Drawbar power: 9.34hp
Transmission: Three-speed
Speeds: 1.3-6.8mph (2.1-10.9km/h)
Fuel consumption: 7.32hp/hr per gallon
Weight: 2,710lb (1,232kg)

You are looking at the most influential tractor of all time. The Fordson F was simpler, lighter and easier to understand than any rival, not to mention cheaper. In the 21st century, it's difficult to visualise exactly what effect it had, but imagine the impact of a new home PC in 2002: easy to use even for first-timers, and one-third the price of comparable machines. Something from a household name that millions of people already knew and trusted.

That was the Fordson F, whose biggest achievement wasn't in its technical

advances (though they were impressive enough) but that it persuaded thousands of small farmers to take the plunge and buy their first tractor. This was mechanised farming for the masses. Henry Ford had been experimenting with tractors for ten years by the time the Model F went on sale – the first prototypes were little more than Model T cars minus bodywork and with tractor-type steel wheels, but the production tractor was quite different. It did away with a conventional chassis, using the engine and transmission as stressed members – that made it lighter than any rival, plus cheaper to make. This in turn gave it a better power to weight ratio from its simple 251ci four-cylinder engine, and easier to handle. In a typically Ford move, the F was built short enough to be parked sideways on a rail wagon, to minimise shipping costs. But this attention to detail meant a price of just $795, ready to roll. And as a farming recession bit in the early 1920s, Ford responded with dramatic price cuts – at one point, a new Fordson cost just $230.

The F wasn't perfect, but over half a million were sold, transforming American farming. They're still part of farming folklore.

Below: One of the most significant tractors ever made, the Fordson F.

Fordson F (1929)

USA/UK

Specifications for Model F petrol (1930)
Engine: Water-cooled, four-cylinder
Bore x stroke: 4.13 x 5.0in (103 x 125mm)
Capacity: 268ci (4,181cc)
PTO power: 29.1hp
Drawbar power: 15.5hp
Transmission: Three-speed
Speeds: 2.1-7.8mph (3.4-12.5km/h)
Fuel consumption: 9.53hp/hr per gallon
Weight: 3,820lb (1,736kg)

By 1924, only six years after launch, the adverts claimed that over 75 per cent of tractors on American farms were Fordsons – they were probably right, too. But through the 1920s there weren't any major improvements to the tractor – Henry preferred to concentrate on turning out the basic unit at the lowest possible cost. Still, after-market manufacturers began to offer a huge range of adaptations from the middle of the decade. There were graders to suit the Fordson, crawler conversions, industrial Fordsons with solid rubber tyres, golf course mowers and pavement sweepers. But Henry Ford seemed to have lost interest in his tractor – he had other things on his mind. The Model T's long-awaited replacement, the A, was ready to roll, which meant a clearing of production decks for the entire company. With all attention focused on the new car, there was no time to develop the Fordson, so production was shifted to Cork in Ireland, and a little later on to Ford's giant new plant at Dagenham, England.

Above: Henry Ford hardly changed his tractor in 22 years.

The tractor seen here is an Irish Fordson, fitted with the big rear mudguards (sometimes called 'orchard fenders') which had the added benefit of preventing the machine rearing up on its rear wheels and falling over backwards. The short wheelbase Fordson was prone to this trouble, which in the days before safety cabs could be fatal for the driver.

Below: Orchard fenders could prevent the Fordson tipping over backwards.

Fordson Model N

USA/UK

Specifications for Model N petrol (1929)
Engine: Water-cooled, four-cylinder
Bore: 4.13in x 5.0in (103 x 125mm)
Capacity: 267ci (4,181cc)
PTO power: 29hp
Transmission: Three-speed
Weight: 3,600lb (1,636kg)

By the late 1920s, it was clear that the Fordson F needed a serious update. The six-month hiatus in production while all the tooling was shipped to Ireland proved an ideal time to make some changes. The result was the Model N. Its engine was bigger than that of the F, thanks to a 0.125in increase in bore size, and rated speed was boosted from 1,000 to 1,100rpm, giving 23hp on kerosene, another six on gasoline, according to Nebraska. The Model T's somewhat basic ignition system was finally ditched in favour of a conventional high-tension magneto. There were cast

front wheels (heavier but stronger than the spoked items they replaced) and the heavier front axle had a downward bend in the middle. There was also a bigger air cleaner and the characteristic orchard fenders were standard. It's just as well the Model N had signficiantly more power than the F, for it also weighed nearly half a ton more.

Sadly, the Irish production plant was not a success. Cork was out on a limb, miles from the big tractor market in England and the London headquarters – raw materials were expensive there, too. To make matters worse, within a couple of years of the plant opening, the worldwide economic slump hit the demand for tractors. The answer was to pack up the machine tools and shift production again, this time to Ford's new plant at Dagenham on the outskirts of London. That was complete by early 1933, and the Fordson N emerged from its new home in a new coat of blue paint and with conventional mudguards. It stayed in production at Dagenham until 1945.

Below: The N was a slightly more sophisticated update on the original F.

Fordson Model N Water Wash (1937)

USA/UK

Specifications for Model N
Engine: Water-cooled, four-cylinder
Bore: 4.13in x 5.0in (103 x 125mm)
Capacity: 267ci (4,181cc)
PTO power: 29hp
Transmission: Three-speed
Weight: 3,600lb (1,636kg)

By the late 1930s, even the Fordson N was starting to look old-fashioned. Although the N was a useful update, the basic format had gone unchanged for 20 years. That had been some attempts to upgrade the N: pneumatic rubber tyres, lights and a starter were now optional, and a rear-mounted PTO was available too. But none of these addressed the basic problem – that the Fordson was out of date.

In England, where the Fordson was now built, it was still very popular, but in America, market share had dwindled to just 5 per cent. The real problem for American farmers was that the Fordson was little use for row-crop work. International had shown the way with its all-purpose Farmall in the mid-1920s – this had a wide rear track and two front wheels set close together, plus extra ground clearance, which made it ideal for row-crops. Ford introduced a belated competitor in 1937, the All-Around, which had the Farmall's tricycle-type layout. It was at least a step in the right direction, and the Sweigard brothers of Halifax, Pennsylvania, certainly liked theirs, according to a Fordson advert: 'We have had

Fordson N Roadless

USA/UK

Engine: Water-cooled, four-cylinder
Bore: 4.13in x 5.0in (103 x 125mm)
Capacity: 267ci (4,181cc)
PTO power: 29hp
Transmission: Three-speed
Weight: 3,600lb (1,636kg) plus track bogies

Sometimes even tractors need help with traction, and in the days before four-wheel-drive became commonplace the answer was a half-track conversion by Roadless. Fordson N's, such as the one pictured here, were particularly popular conversions in England. Roadless was the brainchild of one Lieutenant Henry Johnson, whose early career read like a Boy's Own adventure. When the Boer War between England and South Africa broke out in 1899, this young engineer immediately volunteered for duty. Rejected for defective eyesight, he worked his own passage there by cattle boat, and ended up in Cape Town working for the British Army – here he had the chance to study the use of big heavy vehicles off-road. A spell in India soon afterwards saw Johnson working for steam engine manufacturer Fowler, once again pondering on the off-road traction problems.

Back at home during World War I, he spent much of his time working on tank development, including rubber tracks and a spring and cable suspension system. After the war, he set up in business as Roadless Traction Ltd, replacing the rear wheels on Foden and Sentinel steam lorries with tank-like tracks and bogies. Motor lorries followed, but it wasn't until the late 1920s that Johnson

Above: Rubber tyres couldn't disguise the age of the N's basic design.

our new All-Around Fordson about seven months, and we believe it to be the best and most economical tractor of its size on the market today. This tractor takes about a gallon and a half of gasoline an hour pulling two 14in bottom plows. We do not need to add any oil between changing. We have never had any trouble starting it, for it always starts with the first turn of the crank.'
But not everyone agreed, and the All-Around's sales didn't make a huge impact on Fordson's US market share. What it needed was something genuinely new.

Above: Half-track conversion by Roadless was a popular addition.

turned his attention to the tractor market. A Roadless converted Fordson N was demonstrated on Margate beach in 1930, after which this became a very popular conversion, especially once it was offically approved by Fordson.

Fordson E27N Roadless

USA/UK

Engine: Water-cooled, four-cylinder
Bore x stroke: 4.13in x 5.00in (103 x 125mm)
Capacity: 267ci (4,184cc)
Power: 27hp @ 1,200rpm
Transmission: Three-speed

Roadless converted tractors weren't just used for agriculture. They were ideal for forestry and hauling lifeboats, while during World War II some were used by the RAF to refuel and tow planes. Nor were they all Fordsons. Many Case tractors, including the C, L and LH, were converted by Roadless, whose conversions retained conventional steering wheels. The company also built experimental half-track tractors for a whole range of manufacturers: McLaren, Lanz, Bolinder-Munktell, and Allis-Chalmers all had prototypes built by Henry Johnson's company. During the war, Massey-Harris and Oliver machines were converted as well.

But in England, Roadless was always associated with Ford, a link that was maintained right up to the 1980s. So just as the Roadless N had been a popular conversion, so was its successor the E27N. The conversion was even awarded a silver medal at the 1948 Royal Show.

Once again, it was Fordson approved, and several different versions of the half-tracked E27N were exported. Mindful of the needs of American farmers, Roadless developed a row-crop tricycle conversion of the E27N Major, which stayed in production until 1964.

Fordson Major Roadless

USA/UK

Engine: Water-cooled, four-cylinder
Bore x stroke: 3.74in x 4.52in (94 x 113mm)
Capacity: 199ci (3,104cc)
Rated speed: 1,600rpm
Transmission: Six-speed
Weight: 5,100lb (2,318kg)

Successful though the Roadless half-track conversions were, in the early 1950s Fordson sought to diversify into full-track and four-wheel-drive tractors. The Roadless J17 was one such, a full-track conversion of the standard Diesel Major. The company had already produced a full-track tractor by then – the Model E was an E27N-based machine, with the choice of Fordson TVO or diesel power, and an output of 40hp. Unlike the Roadless half-tracks, these were steered by twin levers, not a steering wheel.

At the same time the company linked up with Selene in Italy, where Dr Segre-Amar was converting Fordson tractors to four-wheel-drive using a transfer box and war surplus GMC truck front axles. This led to Roadless building the same conversion in Britain. It was such a success that the firm went on to covert the Power Major, Super Major and even the smaller Dexta to four-wheel-drive. And, as before, Roadless wasn't exclusively tied to one manufacturer, and also converted an International tractor to 4x4 – the B-450 was in production right up to 1970. Meanwhile, there were 4x4 conversions of the Ford 2000, 3000, 4000 and 5000. Sadly, Roadless did not survive the 1980s – most tractor makers were now producing their own 4x4s, straight off the production line, so the market for specialist conversions was gone.

Above: In the UK, the Roadless was linked with Ford for decades.

Above: Roadless also made full-track conversions, here on a Fordson Major.

Fordson E27N Petrol

USA/UK

Engine: Water-cooled, four-cylinder
Bore x stroke: 4.13in x 5.00in (103 x 125mm)
Capacity: 267ci (4,184cc)
Power: 27hp @ 1,200rpm
Transmission: Three-speed
Weight: 4,000lb (1,818kg)

In the late 1930s, the Fordson N was considered outdated, yet the E27N which succeeded it in 1945 in England was yet another update of the same basic unit. Despite which, it sold respectably well until replaced by the New Major in 1952. How did they do it?

In England, farms were smaller and tractor competition less intense. Also, the Fordson had developed a very good reputation, not just for service in the fields, but thanks to its part in the war effort. One hundred and fifty thousand Fordson Ns had been built at Dagenham during World War II. But in the difficult economic conditions of 1945, there just were not the time and materials to

come up with an all-new tractor, or to build the Ford-Ferguson – it would have to be an update on the N.

So E27N used the venerable Fordson gasoline engine – now up to 267ci and rated at 1,200rpm. 'E' incidentally, stood for 'English', '27' was the horsepower and 'N' made the lineage clear. A water pump was added, but the four-cylinder motor still used splash lubrication. Still, it was powerful enough to make the E27N a three-plough machine, thanks to a new bevel-drive rear axle – the old worm-drive couldn't have taken the extra load. The whole thing weighed 4,000lb (a far cry from Henry Ford's lightweight original) though it did at least have the option of a hydraulic implement lift. With no draft control, the latter did not infringe Harry Ferguson's patents – although in theory Ford and Ferguson were allies, this didn't include the English end of the operation, and relations were deteriorating fast.

Below: Fordson E27N was the final update on the N, with roots in the F.

Fordson E27N Diesel

USA/UK

Engine: Water-cooled, six-cylinder
Bore x stroke: 3.50in x 5.00in (88 x 125mm)
Capacity: 288ci (4,493cc)
Power: 45hp @ 1,500rpm
Transmission: Three-speed
Weight: 4,500lb (2,045kg)

British farmers might have liked the standard E27N, but it had its drawbacks. The elderly motor had been pushed to its limits to keep up with the competition, now in high compression, higher revving form. Now pulling a three-bottom plough behind the E27N, service life was limited. Worse still, overhauling was complicated by the lack of cylinder sleeves – a worn-out engine had to be rebored. But the British had their own solution, a modern diesel.

Diesel engines were already far advanced in Britain, and Frank Perkins had already converted his own Fordson to diesel power. When Ford got wind of

this, they sent two more tractors to him for conversion, and were so impressed with the result that it went into production. Fitting the Perkins P6 six-cylinder diesel transformed the E27N. It weighed 500lb (227kg) more than the standard model, but produced a healthy 45hp at 1,500rpm. Here was a modern affordable diesel tractor that proved to the average farmer that diesel power needn't be complex or expensive. The E27N's original transmission and clutch seemed happy with all that extra power, though the rear axle was strengthened. Whatever, the Perkins-powered E27N (recognisable by the four-rings Perkins badge on the radiator) was a real success, with 23,000 being built before production ended in 1952. It also led to conversions of older Fordson Ns, using both the P6 engine and its four-cylinder counterpart, the P4. When the E27N was replaced in 1952, there was no question that the New Major would have a diesel option from the start.

Below: A Perkins six-cylinder diesel transformed the E27N's performance.

Fordson Power Major

USA/UK

Specifications for Power Major diesel (1959)
Engine: Water-cooled, four-cylinder
Bore x stroke: 3.94 x 4.52in (99 x 113mm)
Capacity: 220ci (3,432cc)
PTO power: 47.7hp @ 1,700rpm
Drawbar power: 42.6hp
Transmission: Six-speed
Speeds: 1.9mph-13.8mph (3.0-22.1km/h)
Fuel consumption: 14.2hp/hr per gallon
Weight: 5,445lb (2,475kg)

With diesel power proven, for the New Major of 1952 Ford developed its own, with TVO and gasoline versions all based on the same cylinder block and crank. The latter two were of 199ci (3.1 litres), the diesel 220ci (3.4 litres), though all were rated at 35hp. Larger and heavier than the E27N, the New Major weighed over 5,000lb (2,273kg), and all versions had a two-range shifter in addition to the

standard three-speed gearbox, giving six forward speeds in all. There were Lucas 12-volt electrics and a hydraulic three-point hitch, but still no live hydraulics or draft control.

The Power Major pictured here was an update of 1959, which brought a significant 22 per cent power increase to the diesel. This came courtesy of a modified fuel injection system, camshaft, cylinder-head and rocker arms. The differential was uprated as well, to cope, and the result was a genuine five-plough machine. There were other technical improvements too: live power take-off and optional power steering, while buyers could choose between standard, tricycle and industrial versions. The Super Major and New Performance Super Major which followed were the same tractor with disc brakes, a differential lock and draft control – the Ferguson situation had evidently moved on. All these tractors were built at Dagenham, but the Super Major was sold through US dealers as the 5000 Diesel up to 1964.

Below: Ford developed its own diesel engine for the Power Major.

Friday 048

USA

Specifications for Friday 048 (1948)
Engine: Water-cooled, six-cylinder
Capacity: 218ci (3,401cc)
Transmission: Ten-speed

The Friday tractor took its name from its proprietor, David Friday. In 1940, Friday, of Hartford, Michigan, had taken over a tractor company with the unlikely name of Love. This was no mass-production enterprise, and Mr Friday reportedly made two tractors during 1940, based on the 1939 Love.

Any plans for a new range of tractors had to be shelved with the outset of war, and Friday, according to *Prairie Farmer* magazine, contented himself with making and servicing Love tractors. These were increasingly made from new parts, not salvaged parts, though in 1942 David Friday did build what he called

the Weber Special, which was indeed made up of salvaged components. He also built two machines powered by Ford flathead V8 engines, and badged them Fridays.

But it wasn't until 1948 that the Friday tractor formally came into being. Not only were the Love orchard tractors renamed Fridays that year, but the first new Friday was unveiled. Powered by a Chrysler-Dodge gasoline six-cylinder engine of 218ci (3,401cc), the new Friday 048 looked the part of a modern, streamlined tractor. It was a three-plough machine with a ten forward, two reverse transmission. It was also one of the fastest tractors of its time (if not all time) with 60mph claimed. The Friday stayed in production until 1948.

Below: Friday claimed a frightening 60mph for its streamlined 048.

Gibson

USA

Specifications for Gibson Model I petrol
Engine: Water-cooled, six-cylinder
Bore x stroke: 3.4 x 4.1in (85 x 103mm)
Capacity: 149ci (2,459cc)
PTO power: 39.5hp @ 1,800rpm
Drawbar power: 29.1hp @ 1,800rpm
Transmission: Four-speed
Speeds: 2.0-14.7mph (3.2-23.5km/h)
Fuel consumption: 10.0hp/hr per gallon
Weight: 4,512lb (2,051kg)

They don't come much rarer and shorter-lived than the Gibson. The company was formed in March 1946 by Wilbur Gibson, as an offshoot from his father's railcar business, which had already begun experimenting with tractors. A new factory was built at Longmount, near Denver, and production began of the little single-cylinder Model A (later D), using a 6hp Wisconsin air-cooled engine. It

had a three-speed transmssion and two independent rear brakes, and measured a flyweight 875lb. All these smallest Gibsons were steered by a lever, though a conventional steering wheel was optional.

Bigger tractors soon followed, notably the two-cylinder Super D2 and E models, and full-size H, which started production in 1948. The latter used a four-cylinder Hercules engine, rated at 25 belt hp. The I model shown here was the biggest tractor Gibson ever made, powered by a six-cylinder Hercules with 40hp at the belt. It was rated as two- or three-plough and weighed around 4,000lb. Three versions were offered: I with tricycle front end, IFS with fixed front axle and IFA with adjustable front axle. Not many Is or Hs were made – probably fewer than 500 each – but over 50,000 of the smaller Gibsons were, according to one estimate. But the company was already busy building forklifts for the US Navy – tractor production ended before Gibson was sold to Helene Curtis Industries in 1952.

Below: The six-cylinder Model I was Gibson's largest.

Graham Bradley Model 32hp

USA

Specifications for Graham Bradley 503.103 petrol
Engine: Water-cooled, six-cylinder
Bore x stroke: 3.25 x 4.38in (81 x 110mm)
Capacity: 218ci (3,399cc)
PTO power: 27hp @ 1,500rpm
Drawbar power: 20hp
Transmission: Four-speed
Speeds: 2.8-19.8mph (4.5-31.7km/h)
Fuel consumption: 10.0hp/hr per gallon
Weight: 4,955lb (2,230kg)

The Graham Bradley was a short-lived tractor, in production for only three years. The Graham brothers had actually started out as farmers, before branching into commercial bodywork and making their first complete truck in 1919. They sold

out to Dodge in 1926, then plunged into the car business, buying up the Paige-Detroit car factory in Dearborn, and making Graham-Paige cars – they became plain Grahams in 1931. But these were hard times for small car manufacturers, and despite selling over 70,000 cars in 1928, fewer than 20,000 rolled off the line in 1936.

The answer appeared to be a tractor. The brothers had already experimented with fitting their six-cylinder 32hp car engine to a tractor, and in 1937 the Graham Bradley tractor was born. It used a 32hp six-cylinder engine, a four-speed gearbox and electric start. Unfortunately, it was not a success, and production ended in 1940/41 before the Grahams sold out altogether to Joseph Frazer in 1944. The brothers went on to make farm machinery for a while, but finally went back to what they knew best – farming.

Below: 1930s Art Deco elegance for the six-cylinder Graham Bradley.

Gray

UK

Specifications for Gray 18-36
Engine: Water-cooled, four-cylinder
Bore x stroke: 4.75 x 6.75in (119 x 169mm)
Capacity: 478ci (7,457cc)
PTO power: 32.2hp @ 950rpm
Drawbar power: 19.5hp
Transmission: Two-speed
Speeds: 2.25 and 2.75mph (3.6 and 4.4km/h)
Fuel consumption: 6.67hp/hr per gallon
Weight: 6,500lb (2,925kg)

The Gray was an unusual tractor indeed, for its appearance, if nothing else. All the mechancial parts were covered by a large sheet of corrugated iron, instead of the usual enveloping bodywork. Launched in 1914, it sought maximum traction by forsaking twin rear wheels in favour of a 54-inch wide

steel roller. This offered the promise of lower ground pressure and being less prone to bogging down. The roller was driven by two exposed roller chains, one each side, each one connecting a system of straight-cut gears to a massive sprocket on each side of the roller. The gearbox was a conventional two-speed however.

All this was fine when moving in a straight line, but the two drive chains operated effectively as two wheels, and there were no differential to aportion power for cornering. So turning at headlands could be difficult on the Gray. A complicated steering system used many rods and levers, and to adjust the fanbelt, you had to move the radiator. At the Lincoln Tractor Trials of 1919, in which many early tractors were put to practical test, the Gray was considered too heavy and expensive. In England, it cost £600.

Below: Somewhere under that corrugated 'roof' is a Gray 18-36. The 54-inch rear roller was intended to improve traction, but made steering difficult.

Hanomag D52

GERMANY

Specifications for Hanomag D52 (1930)
Engine: Water-cooled, four-cylinder diesel
Capacity: 333ci (5.2 litres)
Power: 40hp @ 1,100rpm

Like its compatriot Deutz, the German firm of Hanomag was a pioneer of diesel tractors. While Deutz could claim to have built the first diesel machine in 1926, Hanomag countered that it was first with full diesels (as opposed to semi-diesels). But Hanomag didn't just make tractors. Founded in 1835, it started out as a supplier of steam locomotives and ships: trucks were made from 1905 and cars from 1924.

The first tractors were big motor ploughs from 1912, using Baer, Kamper or

Körting engines. Crawler and wheeled tractors followed after World War I, initially with four-cylinder gasoline engines of 25hp. It wasn't until 1930 that Hanomag's first diesel appeared, after two years of work by chief engineer Lazar Schargorodsky. The D52 used a 333ci (5.2 litres) four-cylinder unit producing 40hp at 1,100rpm. A 545ci (8.5-litre) six-cylinder version followed in 1936. As well as these agricultural tractors, Hanomag reflected its wide product base with the Gigant road tractors (built since the 1920s, and popular as fairground transport right up to the '60s). The company was heavily involved in the German war effort through the late 1930s and 40s, building the well known half-track armoured personnel carrier, over 14,000 of which were built.

Below: Hanomag's four-cylinder D52 was an early four-stroke diesel tractor.

Hanomag '9.430s'

GERMANY

Specifications for Hanomag R22 (1951)
Engine: Water-cooled, three-cylinder diesel
Capacity: 135ci (2,099cc)
Power: 22hp
Transmission: Five-speed (eight-speed option)
Speeds: 2.3-11.4mph (3.7-18.2km/h)
Weight: 1,520kg

Hanomag was one of the few German tractor manufacturers to survive World War II. It continued making the lightweight RL25, which had originally been introduced in 1937, in three- or four-speed form. That was replaced in 1949 by the R25, complete with hydraulics, and three years later the company celebrated completing its 75,000th tractor. In that year (1952) the range spanned 16 to 55hp, with crawlers of up to 90hp.

These were conventional four-stroke diesels of course, but in the 1950s, the company began offering two-strokes as well – the little 12hp single-cylinder R12 was unveiled in 1953, and a twin-cylinder version soon followed. The Hanomag two-strokes weren't without their problems, but helped to boost production to a total of 100,000 by 1954 – another 50,000 were built over the next two years.

Exports helped these figures along, as did a production plant in Argentina. By the early 1960s, Hanomag's two-stroke era was over, and the range spanned from the 25hp Perfekt 300 to the 60hp R460. A completely new range was launched in 1967, using direct injection diesels, three- or four-range three-speed transmissions and a four-wheel-drive option. But sales were disappointing, and tractor production ceased in 1971.

Below: This two-seater 9.430 Hanomag dates from the 1950s.

Happy Farmer

USA

Specifications for La Crosse Model G (1920)
Engine: Water-cooled, twin-cylinder
Bore x stroke: 6.0 x 7.0in (150 x 175mm)
Capacity: 396ci (6,172cc)
Power: 24.2hp @ 900rpm
Fuel consumption: 6.56hp/hr per gall
Weight: 4,670lb (2,102kg)

Happy Farmer – how could any tractor fail with a name like that? Unfortunately, this Minneapolis tractor didn't last very long at all, despite its happy logo, complete with grinning John Wayne lookalike features.

This short-lived venture was established in 1915, the idea being to provide a tractor for implement makers La Crosse to sell against the highly successful

Waterloo Boy. The latter of course, was the first successful gasoline tractor, later bought up by John Deere to provide a quick entry into the tractor market. To this end, Happy Farmer was soon reorganised along with the Start-Rite Engine Company, a firm which had already built many tractors. Together, they formed the La Crosse Tractor Company. The tractor itself was a simple, basic machine, using a twin-cylinder water-cooled engine, mounted with the cylinders horizontal. Unusually, it had pushrod operated overhead valves, instead of the more usual valve-in-head layout, though all the valve gear was exposed. A similar (but not identical) tractor, the La Crosse Model G, was tested at Nebraska in 1920, but the Happy Farmer ceased production in the early 1920s.

Below: Happily named, but the Happy Farmer lasted only a few years.

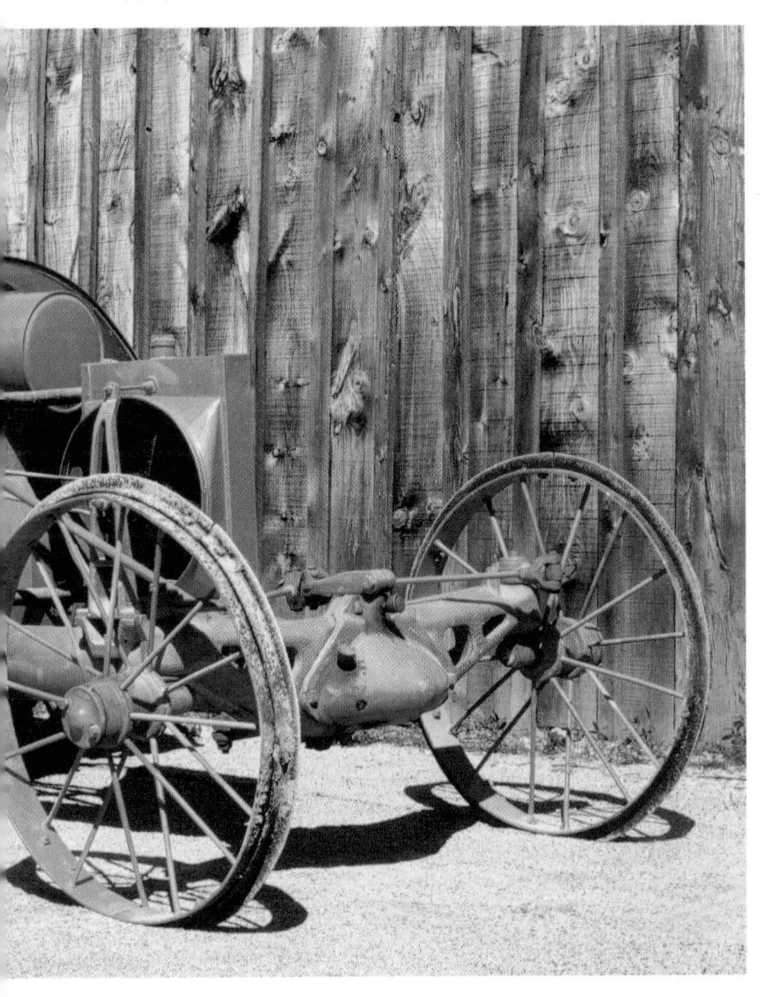

Hart-Parr

USA

Hart-Parr was one of the pioneers of tractor building. In fact, building prototype tractors in 1902, and going into production the following year, it was arguably the first-ever tractor manufacturer, helping to kick-start an entire industry. Not for nothing was the company slogan 'Founders of the Tractor Industry.' Over 25 years, Charles Hart and Charles Parr built up a successful business, first with massive models like the 30-60, later with the more conventional 12-24. Hart-

Parr's sales manager W. H. Williams was even said to have coined the term 'tractor'. But by the company wasn't selling enough tractors to survive on its own, and in 1929, merged with three other companies, including Oliver. The Hart-Parr name was an early casualty.

Below: Hart-Parr was probably the world's first tractor manufacturer.

Hart-Parr 12-24

USA

Specifications for 12-24 (1924)
Engine: Water-cooled, two-cylinder
Bore x stroke: 5.5 x 6.5in (138 x 163mm)
Capacity: 309ci (4,820cc)
PTO power: 27.0hp @ 800rpm
Drawbar power: 17.0hp
Fuel consumption: 9.44hp/hr per gallon
Weight: 4,675lb (2,125kg)

Hart-Parr brought three established tractors to the 1929 merger – the 12-24, 18-36 and 28-50. The smallest one had its origins in the 12-25 of 1918, the machine which drove Messrs Hart and Parr to eventually leave the company. Their partner Charles Ellis had realised that a smaller, lighter tractor was what the market wanted – this was before the Fordson had its major impact – but it wasn't what the founders wanted.

Ellis won through, and his legacy was a conventional, competitive tractor which helped establish the classic format of radiator in front, then engine surmounted by fuel tank and driver's seat nestling between two big rear wheels. It was a twin-cylinder machine with bore and stroke of 5.5 (some say 5.75 in by 6.5in. Turning over at 850rpm, it drove through a two-speed transmission and the whole machine weighed 4,675lb (2,125kg). Although rated as a 12-25, the Nebraska tests revealed this to be conservative in the extreme, with 17 drawbar hp, and 27 at the belt.

Right: Despite a rating of 12-24, H-P's smallest produced 17/27hp!

Hart-Parr 28-50

USA

Specifications for 28-50 (1927)
Engine: Water-cooled, four-cylinder
Bore x stroke: 5.75 x 6.5in (144 x 163mm)
Capacity: 675ci (10,530cc)
PTO power: 64.6hp @ 800rpm
Drawbar power: 43.6hp
Transmission: Two-speed
Speeds: 2.2mph & 3.6mph (3.5 & 5.8km/h)
Weight: 10,394lb (4,725kg)

For its time, this was a true monster of the fields. The Hart-Parr 28-50 weighed nearly five tons and recorded a pull of over 7,000lb in the ever-present Nabraskan tests. H-P already had a scaled-up version of the 12-24, the 18-36, a four-plough machine that used two massive cylinders, each measuring 6.75in x 7.0in (169 x 175mm). Competing against contemporaries like the John Deere Model D and Huber 25-50, it cost $1,350 and was claimed to plough an acre per hour.

Its big brother was another step up. Bigger-still cylinders would probably have caused vibration problems, so Hart-Parr took the novel step of powering its biggest tractor yet with two 12-24 engines, placed side by side. They gave enough power to pull six 14in ploughs or run a 36in threshing machine. And all for just over $2,000, which included mudguards. Of course, it was heavy – fresh from the factory, the 28-50 weighed 8,600lb (3,909kg), and when working that could be more than 10,000lb (4,545kg). As with the 12-24, its actual power was way in advance of the conservative rating – nearly 65hp was a lot for 1927.

***Above:* Twin-engined 28-50 used two 12-24 power units, side by side.**

HSCS

HUNGARY

Specifications for HSCS K40 'Steel Horse'
Engine: Single-cylinder, semi-diesel
Power: Not known
Transmission: Not known
Weight: 7,500lb (3,409kg)

HSCS hailed from Hungary, and was active in tractor making early on, building its first machine in 1923. It started making gasoline engines four years earlier, and was formed when the Clayton and Shuttleworth company pulled out of the Hungarian steam traction engine market. Hence the name, an acronym for Hofherr, Schrantz, Clayton and Shuttleworth.

That first 1923 tractor used a single-cylinder gasoline engine, but HSCS soon switched over to a semi-diesel motor, to which it remained faithful for many years. The semi-diesel met with limited acceptance in the USA and

Britain, though the concept was popular in much of Europe – Landini, for example, used a semi-diesel right up to 1957. A semi-diesel is mechanically simple and will run on almost any type of fuel, including waste oil, though it is rather crude compared to multi-cylinder four-stroke diesels.

HSCS's motor was rated at 14hp, intended for ploughing and as a stationary power source, and derivatives continued to power a range of wheeled and tracked tractors through the next few decades. In 1951, under the Communist regime, HSCS was renamed (appropriately enough) Red Star tractors, using the DUTRA brand name after 1960. The company later built Steiger tractors under licence, named RABA-Steigers.

Below: Most HSCS tractors used the two-stroke semi-diesel layout, similar to that of Lanz, Landini, SFV and others.

Huber 15/30

USA

Specifications for 15-30 (1921)
Engine: Water-cooled, four-cylinder
Bore x stroke: 4.5 x 6.0in (113 x 150mm)
Capacity: 382ci (5,959cc)
PTO power: 44.7hp @ 1,000rpm
Drawbar power: 26.9hp
Transmission: Two-speed
Speeds: 2.7mph & 4.2mph (4.3 & 6.7km/h)
Fuel consumption: 9.23hp/hr per gallon
Weight: 6,090lb (2,741kg)

Huber of Marion, Ohio, was one of the true tractor pioneers, and also one of the most successful in the early years. The company was a well-established maker of steam traction engines and buitl its first internal combustion engined machine in 1898. Rather than design something new from scratch, Huber

simply fitted a proprietory Van Duzen engine and bolted it into one of its existing steam engine chassis and transmssions.

It was a success, and 30 were made in the first year, while Huber took over Van Duzen to ensure its engine supply. Actually, Huber would buy-in engines from outside right through to the 1930s, from MidWest, Waukesha and Stearns, among others. Whoever it bought engines from, Huber was to be a prominent part of the tractor industry for the next 40 years. It made road rollers as well as tractors, the first i-c engined roller appearing in 1923 – Huber publicity material claimed this had revolutionised road building in the USA.

The 15-30 pictured here was Huber's midrange tractor, using a 382ci (5,959cc) MidWest four-cylinder engine. In fact, the company was evidently fond of the four-cylinder layout, and every single Huber tested by the University of Nebraska had four-cylinders.

Below: A distinctive layout for this Huber, with transverse four-cylinder engine.

Huber 40/62

USA

Specifications for 40-62 (1927)
Engine: Water-cooled, four-cylinder
Bore x stroke: 5.5 x 6.5in (138 x 163mm)
Capacity: 617ci (9,625cc)
PTO power: 69.8hp @ 1,100rpm
Drawbar power: 40.6hp
Transmission: Two-speed
Speeds: 2.4mph and 3.4mph (3.8 and 5.4km/h)
Fuel consumption: 6.58hp/hr per gallon
Weight: 9,910lb (4,460kg)

Huber made a whole range of tractors, from the 12/25 up to this one, the 40-62. It had actually started life as a 25-50, until tested by Nebraska, who found that it had 60 per cent more drawbar power than Huber claimed, and 40 per cent more at the brake! A pity that more manufacturers weren't so modest in their predictions.

In any case, the 40-62 (as it was swiftly renamed after Nebraska's revelations

in June 1927) used a big Stearns four-cylinder engine of 617ci. It was a valve-in-head unit, rated at 1,100rpm and consuming fuel at the rate of 6.45 gallons per hour – this was before Nebraska came up with its hp/hr per gallon formula. But if the 40-62 performed well, the same could not be said of the smallest Huber, the Light Four 12-25, which went through the Nebraska treadmill in May/June 1920. During 45 hours of running time, it was plagued with minor faults. The carburettor had to be replaced and the magneto adjusted, then readjusted. The fan belt had to be replaced, as did the water tank, which was leaking. Even the valves had to be reground!

The last Hubers to be tested – an LC and streamlined B in late 1937 – weren't without their problems either. The LC slipped out of gear several times, its fuel tank leaked and the oil pressure gauge line broke. A pity, because the B, one of Huber's last tractors, looked modern and up to the minute.

Below: Huber's biggest tractor started life as a 25/50, until independent tests revealed it to be a 40/62! Engine was a big Stearns four-cylinder.

Hurlimann

SWITZERLAND

Specifications for D100
Engine: Water-cooled, four-cylinder
Power: 45hp @ 1,600rpm
Transmission: Five-speed

Think of the leading tractor producing nations, and Switzerland does not immediately spring to mind. The USA, China and what used to be the USSR, certainly; the bigger nations of Western and Eastern Europe, of course – but Switzerland?

Hans Hurlimann started in business with a single-cylinder gasoline-powered machine, fitted with a power bar grass mower. It was small and manoeuvrable – Hurlimann tractors have always reflected their native terrain. Swiss farms tend to be small and hilly, with little room for the development of big arable crops. So Hurlimanns have been small as well, but well engineered.

One new machine announced in 1939, for example, had a direct injection four-cylinder diesel. The D100, introduced as World War II ended, added a five-speed transmission, PTO, two-speed belt pulley and differential lock. With 45hp at 1,600rpm, it was powerful enough, and also had a usefully low centre of gravity, thanks to smaller than usual rear wheels.

Hurlimann carried on producing tractors in the decades that followed, one example being the 5200 of the 1970s, and it's also worth noting that despite Switzerland's small size, it was not the only tractor maker based there. In Hurlimann's current line-up is the small Prince 435, which uses a 35hp Mitsubishi diesel, four-wheel-drive and a 12-speed transmission. Hydrostatic power steering and fold-down ROPS protection are also standard. In the USA, it costs just under $20,000, which includes a front loader.

Below: One of the few Swiss tractor makers – Hurlimann's 5200 is typical.

International Harvester

USA

Like so many American tractor makers, International Harvester was the result of a merger. In 1902, McCormick Harvesting Machine, the Deering Harvester Company and three smaller firms all joined to form IHC. The pairing of McCormick and Deering was an odd one, for they were fierce rivals, and remained so even after the merger. The two insisted on separate tractor lines, and even separate dealerships in the same town, but this didn't seem to harm IHC, which concentrated on the growing market for small tractors. It was soon the leading US manufacturer.

But International's big break came in 1923 with the Farmall – here was a tractor that could literally do it all, and it was soon outselling the rest of the IHC range put together.

Variations on the Farmall theme followed, plus bigger standard-tread tractors like the W30, which replaced the now elderly 15-30, and there were

crawler versions as well. The famous stylist Raymond Loewy was engaged in 1936, and the result was a new range of modern, streamlined tractors with more attention given to ergonomics – the A and B, while the Farmall was updated as the H and M. But the 1950s were not good for International. Powershifting was a big advance in 1953, but IHC was getting left behind in the horsepower race, and rushed to build a new six-cylinder machine. Too powerful for the Farmall driveline, it had to be recalled and redesigned at great expense. To add salt to the wound, IHC was now outsold by John Deere. Despite more new features in the '60s and '70s, International was weakening. It could not survive on its own, and in 1984 was taken over by Tenneco Corporation – Case-International was born.

Below: IHC built many tractors, but made its fortune with the Farmall.

International Titan

USA

Specifications for 10-20 (1916)
Engine: Water-cooled, two-cylinder
Bore x stroke: 6.25 x 8.0in (156 x 200mm)
Capacity: 490ci (7,644cc)
Drawbar power: 10hp
Belt power: 20hp
Transmission: Two-speed
Speeds: 2.0 and 2.75mph (3.2 and 4.4km/h)
Weight: 6,138lb (2,762kg)

International's two rival factions – McCormick and Deering – demanded two rival tractor lines. So McCormick dealers were given the Mogul line, while the Deering men sold Titans. For the most part, Moguls were designed and built at IHC's Chicago factory, using throttle-governed engines, while Titans were powered by Famous power units (with less reliable governors) and came from Milwaukee. Fortunately, both the 8-16 Mogul of 1914 and 10-20 Titan

(which appeared a year later) were successful.

So successful that IHC suspended production of its big tractors in 1917 to concentrate on the small ones – after all, these relatively light and affordable machines were far more useful to the average farmer than some great behemoth that owed as much to steam traction engine technology as the new internal combustion engine. The plan certainly worked well for IHC, which sold over 17,000 Titans in 1918 alone. Interestingly, some 3,500 were exported to Britain during World War I. Along with the Fordson they hugely significant in introducing UK farmers to the advantages of a modern tractor.

The Titan's twin-cylinder engine started on petrol, then ran on kersene once warmed up. If it ever pinked (pre-ignited) under load, water injection provided a solution. There was no radiator as such; instead, a 35-gallon water tank (looking very much like a steam engine boiler) simply cooled the coolant via its own radiated heat.

Below: IHC made two rival tractor lines – the Titan was for Deering dealers.

International 8-16 USA

Specifications for 8-16 (1920)
Engine: Water-cooled, four-cylinder
Bore x stroke: 4.25 x 5.0in (106 x 125mm)
Capacity: 284ci (4,430cc)
PTO power: 18.5hp @ 1,000rpm
Drawbar power: 11.0hp
Transmission: Three-speed
Speeds: 1.8-4.1mph (2.9-6.6km/h)
Fuel consumption: 7.32hp/hr per gallon
Weight: 3,660lb (1,647kg)

In its early days, International made two tractors rated as 8-16 machines. But although only a few years apart – one was unveiled in 1914, the other in 1917 – they were completely different, underlining the fast pace of tractor development at that time.

The first one, the Mogul 8-16 was a simple two-speed tractor, advanced for its time. But just three years later it was made to look decidedly old-fashioned by its replacement (now sold as an International, not a Mogul). This had a modern four-cylinder engine, running at the high speed of 1,000rpm. The new International had a three-speed transmission and at 3,660lb (1,647kg) was significantly lighter than its predecessor. It also featured a power take-off, a production first for an American tractor, and really the first in the world to be commercially successful. And in another pointer to the future, engine and radiator were covered by a hood, not exposed to the air.

International McCormick-Deering 10-20 USA

Engine: Water-cooled, four-cylinder
Bore x stroke: 4.25 x 5.0in (106 x 125mm)
Capacity: 284ci (4,431cc)
Drawbar power: 10hp
Belt power: 20hp
Transmission: Three-speed
Speeds: 2.0-4.0mph (3.2-6.4km/h)
Weight: 3,945lb (1,775kg)

Although International's 8-16 was advanced in some ways, its separate chassis and chain drive looked crude compared to a Fordson, which was of course much cheaper to buy. International hit back by offering a free plough with every new tractor, enabling it to clear its shipping yards of unsold stock, and introduced a new machine.

The 15-30 of 1921 answered both those criticisms of the 8-16 – it had gear final drive, properly enclosed in an oil bath, and a stressed (not riveted) chassis. Just comparing it with its International predecessor of 1919, shows again just how quickly tractors were developing: the old 15-30 weighed 8,700lb, the new one 5,750lb (2,588kg); it was 13 feet long and cost $2,300 – the new one was just 11 feet long and cost $1,250. Both tractors could do the same amount of work, but the new 15-30 was easier to use, needing less TLC, and had a covered-in engine and driveline. International even put a lifetime guarantee on

Above: Four cylinders and three speeds made the 8-16 quite advanced.

Oddly, despite all these pioneering features, the 8-16 was quite old-fashioned in other ways. For example, it still used a riveted separate chassis – contemporary best thinking was to do away with the chassis altogether and use the engine and transmission as stressed members. Also, the exposed chain and sprocket final drive looked crude next to an enclosed shaft and gear drive. but no matter, this little 8-16, the first to wear the International badge, consolidated the company as a major manufacturer.

Above: Enclosed driveline and lifetime crank guarantee for the 10/20.

the ball-bearing crank, to underline the fact. If the cylinders wore out, they had replaceable liners. This was user convenience at its best. Not surprisingly, International sold over 128,000 15-30s in eight years.

The 10-20 pictured here was simply a smaller version ofthe same thing, with all the same advances, but a 284ci (4,431cc) four-cylinder engine in place of the 15-30's 381ci (5,944cc) unit. It was a huge success – over 216,000 were sold before production ceased in 1942. Incidentally, both these machines were badged as 'McCormick-Deering' because those two rivals, after 20 years under the same roof, still couldn't stomach a corporate name.

International Farmall

USA

Specifications for Farmall F20
Engine: Water-cooled, four-cylinder
Bore x stroke: 3.75 x 5.0in (94 x 125mm)
Capacity: 220ci (3,432cc)
Drawbar power: 16hp @ 1,200rpm
PTO power: 24hp
Transmission: Four-speed
Speeds: 2.25-3.75mph (3.6-6.0km/h)
Weight: 3,950lb (1,778kg)

Most tractor manufacturers have a landmark machine; very few have one that is a landmark for the entire industry. The International Farmall was one such – for the tractor industry, it was just as significant as the Fordson, and probably one of the most important tractors ever built. Until then, tractors were either small and light for cultivation, or heavier and more powerful for drawbar or belt work. The tractor Valhalla was something with over 20hp at the belt, but small, light and agile enough to work the fields without damage. In other words, a tractor that could do it all – the Farmall could.

It was seven years in the making – with remarkable forsight, International engineers had started work on the project back in 1916 – but stll took the market by storm when it finally arrived in 1924. With 16 horsepower at the drawbar and weighing 3,650lb (1,643kg), it was hefty enough to pull a two-bottom plough, while 24 belt horsepower was enough to drive threshers and shredders. On the other hand, it was nimble, able to turn in its own length, while high clearance and wide wheel spacings meant it could drive between rows of cotton or corn without damaging anything.

The odd thing was, IHC management took a while to realise just how revolutionary the Farmall was. They were afraid it would take sales from the superficially similar 10-20 (which had about the same power) and actually held production back for the first year or so – only 200 Farmalls found homes in that first year of production. But then the floodgates of demand opened – over 4,000 Farmalls were sold in 1926, and ten times that in 1930. Tractors would never be the same again.

Below: Breakthrough! Farmall combined power with agility to great effect.

International Farmall F30 USA

Specifications for Farmall F30
Engine: Water-cooled, four-cylinder
Bore x stroke: 4.25 x 5.0in (106 x 125mm)
Capacity: 284ci (4,430cc)
Drawbar power: 20hp @ 1,150rpm
PTO power: 30hp
Transmission: Four-speed
Speeds: 2.0-3.75mph (3.2-6.0km/h)
Weight: 5,300lb (2,385kg)

Given the impact and sales success of the original Farmall, it wasn't long before John Deere, Case, Allis-Chalmers and all the rest began work on their own more powerful row-crop tractors to rival it.

Reflecting the long gestation period of the original Farmall, the 30 per cent bigger F30 didn't arrive until 1931, though really it was little different. The engine was substantially bigger, with an extra half-inch on the bore to give a capacity of 284ci (4,430cc) – the dimensions were the same as the older 10-20, which was still selling steadily. Despite a slightly lower rated speed of 1,150rpm, the result was just over 20hp at the drawbar, and a genuine 30hp at the belt. Perhaps with one eye on the perceived clash between 10-20 and the original Farmall, IHC had already uprated the 15-30 into a 22-36. The first Farmall, incidentally, was now named the Regular, to differentiate it from its big brother, and was given slightly more power and better steering as the F20 in 1932.

International Farmall F12 USA

Engine: Water-cooled, four-cylinder
Bore x stroke: 3.0 x 4.0in (75 x 100mm)
Capacity: 113ci (1,763cc)
Drawbar power: 10hp @ 1,400rpm
PTO power: 15hp
Transmission: Three-speed
Speeds: 2.25-3.75mph (3.6-6.0km/h)
Weight: 2,700lb (1,215kg)

One thing you could say for International – it might take a long time to come up with new tractors (the Farmall took seven years from drawing board to farmer's field) but variations on the theme didn't always take that long. The company was so busy keeping up with demand for the original Farmall that it took another seven years to produce a more powerful version, the F30.

In the meantime, there was nothing for the thousands of small farmers in America, those with only a hundred acres or so to work. What they wanted was a smaller version of the Farmall, and only a year after the F30 appeared they finally got it. Just as the 30hp F30 was simply a bigger brother of the original Farmall, so the F12 was no more nor less than a scaled down version. In fact, it was much smaller than the original – the F30 was 30 per cent bigger, but F12 had a little 113ci engine, less than half the size of that of the first Farmall. With everything on a smaller scale, it came in at a featherweight 2,700lb, which gave its modest 10hp more of a fighting chance. In any case, it was enough for a one-bottom plough, and, just like the bigger Farmalls, this one could fit between rows, and was able to cultivate two rows at a time. One new feature was the

Above: Farmers who wanted a more powerful Farmall could choose the F30.

The F30 was well equipped as standard, with a belt-pulley, solid-rim wheels (rubber tyres later became an option) an adjustable radiator shutter and regular or narrow tread. It was built until 1939, but proved far less popular than the first Farmall – in fact, it missed the point entirely, as a bigger demand lay waiting not for a bigger, more powerful Farmall, but a smaller one.

Above: Mini-Farmall – the one-plough F12; smaller 10hp version of the original.

adjustable width rear axle, altered by sliding the rear wheels along splines.

The very first F12s used Waukesha engines, but from May 1933 IHC's own power unit replaced it. It was uprated slightly into the F14 in 1938 (to 12hp at the drawbar, 15 at the belt) and finally replaced by the Farmall A the year after.

International Farmall W12 USA

Engine: Water-cooled, four-cylinder
Bore x stroke: 3.0 x 4.0in (75 x 100mm)
Capacity: 113ci (1,763cc)
Drawbar power: 10hp @ 1,700rpm
PTO power: 16hp
Transmission: Three-speed
Speeds: 2.14-3.6mph (3.4-5.8km/h)
Weight: 2,900lb (1,305kg)

Although the Farmall was terrifically popular and versatile, there was still plenty of demand for a standard-tread tractor in the 1930s. International responded by updating the venerable 15-30 in 1932. Renamed the W30, it used the Farmall F30's engine, which meant a slight power drop to 20-31 horsepower, but was cheaper to produce.

The W12 was simply a standard tread version of the little Farmall F12. Mechanically, it was very similar to the F12, albeit with rated engine speed increased to 1,700rpm, and 16hp at the belt, rather than 15. There was a pneumatic rubber tyre option, and the W12, which was introduced in 1934, was adaptable to orchard and golf course work, as well as the field. 'What the W12 will Do,' announced a brochure of the day: '1) Will more than pay its way on any diversified farm... 2) Will plough from 4 to 7 acres a day, double-disk from 16 to 30 acres a day... 3) With pneumatic tyres will do most of the hauling, field and roadway... 4) On many farms will advantageously supplement the work of larger tractors.'

International WK40 USA

Engine: Water-cooled, six-cylinder
Bore x stroke: 3.75 x 4.5in (94 x 113mm)
Capacity: 298ci (4,649cc)
Drawbar power: 38hp @ 1,750rpm
PTO power: 53hp
Transmission: Four-speed
Speeds (pneumatic): 2.4-12.0mph (3.8-19.2km/h)
Weight (WD40): 7,550lb (3,398kg)

By the mid-1930s, International Harvester had not built a larger tractor – in the early days, it offered a whole range right up to a 60hp gargantuan weighing 22,000lb (9,900kg), but these were dropped in 1917 after the huge success of the little 8-16 Mogul and 10-20 Titan.

This meant they were missing out on a growing part of the tractor market, especially in the wheatland prairies of the Mid West USA, where farmers needed something rather bigger than even the most powerful Farmall. International's first response was actually a diesel, the industry's first mass-produced diesel tractor. With 52hp at the belt and 37 at the drawbar, the WD40 took IHC up to a new power class. It was also unique in that the diesel engine had to be started on gasoline before it would run on pure diesel. So as well as the usual diesel injectors and pump, it was equipped with carburettor and spark plugs! It was a big four-cylinder engine, at 355ci (5,538cc), with a 14-gallon (63-litre) cooling capacity and a 31-gallon (141-litre) fuel tank. Ready to ship, the complete tractor weighed 7,550lb (3,398kg) and came with steel wheels or pneumatic rubber tyres.

Above: Standard tread axles turned the Farmall F12 into a W12.

It was of course a more limited production tractor than the F12, and 3,622 were built between 1934 and 1938, when it was replaced by the W14. Like the F14, this was no more than a slightly uprated version of the same tractor. Electric lighting and an adjustable drawbar hitch were options, as well as those pneumatic tyres.

Above: WD (diesel), WK (distillate) and WA (gasoline) were IHC's big 1930s tractors.

The WK40 shown here was the distillate version, with very similar power to the WD40 but a completely different, six-cylinder engine. It was actually a truck engine, already built in-house by International, and also came in gasoline form, in which case the tractor was a WA40. If the pneumatic tyres were specified, the four-speed gearbox was given a high top ratio, for a road speed of 12mph.

International TD40

USA

Engine: Water-cooled, four-cylinder
Capacity: 355ci (5,538cc)
Drawbar power: 37hp @ 1,750rpm
PTO power: 52hp

As well as its wheeled tractors, Internaitonal also built a range of crawlers from the 1930s. Badged as 'Tractractors' they were based around the existing 20-, 30- and 40hp machines, though he most popular was the smallest T20 Tractractor. Introduced in 1931, this was small enough to be profitable for farming as well as construction work. It also came in low seat configuration for orchard work.

Top of the range was the TD40 shown here, which used that unique diesel engine which also saw service in the WD40 tractor. Today, we are used to diesels which start instantly, hot or cold, thanks to electronic injection and quick-heat glowplugs. But tractors and crawlers of the 1930s were really pioneering this type of power unit, and cold starting was something that brought different solutions. One was the so-called 'hot bulb' semi-diesel, used by manufacturers such as Marshall and Landini. This worked on the two-stroke principle, having no valves. To start it first thing in the morning, a blowtorch was used to heat up the cylinder-head, after which it would (hopefully) thump into life – once running, the engine's own heat would keep it going. International's four-cylinder four-stroke, by contrast, would start up on gasoline (just like contemporary distillate engines) using its own carburettor, spark plugs and magneto. Once running, the driver

International Farmall A

USA

Engine: Water-cooled, four-cylinder
Bore x stroke: 3.0 x 4.0in (75 x 100mm)
Capacity: 113ci (1,763cc)
Drawbar power: 17.4hp @ 1,400rpm
PTO power: 19.1hp
Transmission: Four-speed
Speeds: 2.3-9.6mph (3.7-15.4km/h)
Weight: 1,870lb (842kg)

The Farmall A was a direct replacement for the popular F12 and F14 – IHC hadn't forgotten that there were plenty of small farmers out there, and they didn't all need (still less could afford) a 50hp super-tractor. Like the F14, the A was a small single-bottom plough machine, though the 113ci overhead valve engine was uprated to give 13 drawbar horsepower, 16 at the belt.

It was designed specifically with cultivating crops in mind. Named the 'Cultivision' the Farmall A's chassis was offset to the left, giving the driver a clear view ahead. There were adjustable-tread front wheels, while the rear tread was adjustable as well, though not on sliding axles.

Introduced in July 1939, the A was remarkably light, at just 1,870lb and one interesting option (available on other Internationals) was high-altitude pistons. These altered the compression ratio to cope with running in higher country, and came in 5,000-feet or 9,000-feet versions.

The A was produced right through World War II, and its best year was actually 1941, when nearly 23,000 were built. At the war's height, only 105 Farmall As left the line, reflecting a factory busy on war work. In 1947, the

Above: IHC made crawlers too, powered by tractor engines.

would switch over to diesel. A complicated solution, but one worthwhile for farmers needing the extra economy of diesel for long days in the fields. Still, if all of that put you off, International offered its biggest crawler in conventional gasoline/kerosene form as well.

Above: Farmall A replaced the first-generation F12 and F14.

smallest Farmall was given hydraulics, as the Super A, which was produced through to 1954.

International Farmall AV

USA

Engine: Water-cooled, four-cylinder
Bore x stroke: 3.0 x 4.0in (75 x 100mm)
Capacity: 113ci (1,763cc)
Drawbar power: 17hp @ 1,400rpm
PTO power: 19hp
Transmission: Four-speed
Speeds: 2.9-12.8mph (4.6-20.5km/h)
Weight: 2,280lb (1,026kg)

Every tractor range has its own language, and in International Harvester terms in 1939 'V' signalled high-crop. So the AV pictured here was simply the high-crop version of the standard A. There were high-crop versions of the bigger H and M Farmalls as well.

Mechanically, the AV was virtually identical to the A, but something more obvious it shared was the styling. In fact, the A and AV were part of a new generation of Internationals, with special attention paid to styling and ergonomics. In the mid-1930s, more than one manufacturer brought in industrial designers from outside the world of tractors. In IHC's case it was Raymond Loewy, who had a good track record in making utilitarian objects look good. He certainly succeeded with the Farmalls, enclosing the fuel tank, steering bolster and radiator in a single streamlined housing. Lowey also styled the wheels to give an impression of strength and lightness, and even hired an orthopedic surgeon to shape the seat.

International W4

USA

Engine: Water-cooled, four-cylinder
Bore x stroke: 3.31 x 4.25in (83 x 106mm)
Capacity: 152ci (2,371cc)
Drawbar power: 22.5hp @ 1,650rpm
PTO power: 25hp
Transmission: Five-speed
Speeds: 2.3-14.0mph (3.7-22.4km/h)
Weight: 3,890lb (1,751kg)

International's W series of tractors, introduced in 1940, were a mixture of the familiar and completely new. The W9 was the new one, which extended the line-up into another power class, with a 335ci four-cylinder engine producing 35hp at the drawbar, 45 on the belt. There was a diesel version too, the WD-9. But W4 wasn't new, simply a standard-tread version of the phenomenally successful Farmall H. So it used exactly the same 152ci four-cylinder engine, in the same state of tune. The only substantial differences were the standard-tread front axle, though the W4 was also slightly lower geared than the Farmall on which it was based, and the W4 was a little heavier. Top speed on the standard pneumatic rubber tyres was 14mph. Reflecting the American market for tractors, which was dominated by row-crop machines, the W4 sold in small numbers compared with the Farmall. In its best year (1951), 3,519 were sold, and by the time production ended three years later 24,377 had rolled off the production lines.

There were of course variations on the same theme, all of them specialised formats compared with the mass-produced, mass selling Farmall. O4 and OS4

Above: High ground clearance turned the Farmall A into an AV.

As well as the AV, there was an industrial version of the standard A, badged as an International A. It was a quirk of history that IHC agricultural tractors were still badged 'McCormick-Deering' and the crawlers 'Tractractors'. The Deering name was finally dropped inthe late 1940s, and McCormick soon after.

Above: IHC's W-series sold in fewer numbers than the row-crop Farmalls.

were designed for orchard work and the industrial I4 came in standard and heavy duty versions. Meanwhile, a slightly larger W4, the Super W4, was produced in 1953-64, with 164ci (2,558cc) motor and live hydraulics.

International W6

USA

Engine: Water-cooled, four-cylinder
Bore x stroke: 3.86 x 5.25in (97 x 131mm)
Capacity: 248ci (6,200cc)
Drawbar power: 33hp @ 1,450rpm
PTO power: 37hp
Transmission: Five-speed
Speeds: 2.3-14.5mph (3.7-23.2km/h)
Weight: 4,830lb (2,174kg)

Just as the W4 was the standard-tread version of the Farmall H, the W6 did the same job for the larger Farmall M. That, of course, had replaced the Farmall F30 back in 1939, and, though not quite as big a seller as the Farmall H, was still hugely popular – over 270,000 of them had been produced by the time production ended in 1952.

Like the H, it combined the new Raymond Lowey styling, with that rounded radiator grille and ergonomic design, with a 248ci (6,200cc) four-cylinder engine in gasoline or distillate form. This was a 33hp three-plough tractor, and until the arrival of the 45hp W9 the biggest offered by International. But according to author P.W. Ertel, it was more nimble and easier to operate than the little F12, yet could do four times the amount of work. Or if you prefer, in the words of International itself: 'Farmall M has easy operation, comfort and all those other refinements that go to make up a tractor that is a delight to use...It will pull three 14- or 16-inch bottom ploughs under harder than average soil conditions at good ploughing

International WD6

USA

Engine: Water-cooled, four-cylinder
Bore x stroke: 3.86 x 5.25in (97 x 131mm)
Capacity: 248ci (3,869cc)
Drawbar power: 31hp @ 1,450rpm
PTO power: 36hp
Transmission: Five-speed
Speeds: 2.3-14.5mph (3.7-23.2km/h)
Weight: 5,250lb (2,363kg)

Slowly but surely, diesel power was becoming a recognised option by American farmers in the late 1930s, but it was taking a while. In Europe diesel development was little faster until after World War II. The outsider might wonder why diesel took so long to catch on. After all, the principle had been around since 1910, and the very strengths of diesel – long life, efficiency and high torque – were ideal for tractor use.

But the bottom line was that diesels cost more to build, and since every farmer was his own businessman, that was an important factor in holding back sales, especially while gasoline fuel was relatively cheap. In the USA, Caterpillar led the way, developing a range of diesels for its crawlers from 1929 onwards, and International produced its own diesel-powered crawler in 1932. Its first wheeled diesel tractor was the WD40 of 1935. That used a hybrid diesel/gasoline motor, equipped with both injectors and a carburettor – it could run on either fuel.

The WD6 shown here was far more conventional, using a straight diesel conversion of International's existing 248ci (3,869cc) four. In fact, it produced

Above: Clean lines and slatted grille marked out the Loewy styled tractors.

speed...In a word, Farmall M has what it takes to deliver satisfactory and economical power under any field conditions.'

Above: Early four-cylinder diesel power was available for this WD6.

almost identical power figures as well – 31/36hp against 33/37hp – and was rated at the same 1,450rpm. The diesel engine added 420lb to the weight of the tractor, though gearing was unchanged. It was cheaper to run of course, but that bottom line ensured that not many farmers went for this early diesel option – the WD6 cost $3,124 to buy in 1951 – over 30 per cent more than the equivalent W6.

International Farmall BMD USA

Specifications for Farmall H
Engine: Water-cooled, four-cylinder
Bore x stroke: 3.3 x 4.25in (83 x 106mm)
Capacity: 152ci (2,371cc)
Drawbar power (gasoline): 19hp @ 1,650rpm
PTO power: 24hp
Transmission: Four-speed
Speeds: 2.6-15.7mph (4.2-25.1km/h)
Weight: 3,725lb (1,676kg)

A year after the Farmall A was unveiled, it was joined by the B. This was simply a two-row version of the same thing, with tricycle front wheels and a long left axle. It also had a centrally mounted chassis and controls, unlike the A's offset arrangement. Rear tread width was adjustable between 64 and 92 inches, but unfortunately this wasn't enough for many farmers. This

wasn't addressed until 1948 when the Farmall C replaced it. Like its predecessor, this used the A's 113ci (1,763cc) four-cylinder engine, but added big rear tread adjustability, on sliding axles. There were lots of variations too, such as the BMD shown here, which used a Perkins L4 diesel engine and was a British-built model.

But to put the Farmall B in context, although 75,000 odd were sold in eight years, it was a low-volume tractor compared to its big brother. The Farmall H, unveiled in 1939, was one of the most successful Internationals ever – over 390,000 were built before production ceased in 1953. This was IHC's mid-range machine, successor to the original Farmall Standard, with a 152ci (2,371cc) overhead valve four and all-up weight of 3,725lb (1,676kg). And, of course, it shared the new generation Raymond Lowey styling with the A, B and M Farmalls.

Below: Farmall B was a wider tread version of the A – here in diesel form.

International 956XL

USA

Engine: Water-cooled, six-cylinder diesel
Bore x stroke: 3.9 x 5.1in (98.4 x 128.5mm)
Capacity: 365ci (5,687cc)
Power: 95hp
Torque: 364Nm
Transmission: 16-speed
Weight: 10,274lb (4,670kg)

International's 956XL and the slightly more powerful 1056XL were something of legends in their own time. In production for ten years (and with an engine in production for twenty) they were above all else solid, reliable pieces of machinery – with proper servicing, the engine could manage 10,000 hours of hard running without needing major work.

These were European tractors, born and bred. The original 946 and 1046 of 1975 were German-built, and powered by the same D358 six-cylinder diesel as the XLs. These were replaced by the French-built 955 and 1055, in 1977, with the XL cab unveiled four years later. The XL 'Control Centre' was a big step forward in cab design, made at IHC's plant in Croix, France, and fitted to the new 'Fieldforce' range of tractors. It had a one-piece frame and was isolated from the chassis and gave a high level of comfort. Quiet too, at 82dB.

A year later, the 956 and 1056 came along as updates, both fitted with the XL cab as standard. Power (up by 5hp) and torque were slightly increased, and the four-wheel-drive versions incorporated a ZF self-locking limited-slip differential. There were few big changes in the tractors' ten-year career: the Sens-O-Draulic system was added in September 1983, and a centre-drive ZF front axle in '86 – that also gave a much improved turning circle of just 176in (4.5m), less than half the original! Both 956 and 1056 were phased out in 1992.

Left: Halfway through the 956XL's life, it acquired a new badge. IHC was taken over by Tenneco – the parent company of Case – in 1984. Thereafter, all tractors were badged Case-Internationals.

International 1256

USA

Specifications for International 1086 (1977)
Engine: Water-cooled, six-cylinder turbo-diesel
Bore x stroke: 4.3 x 4.75in (108 x 119mm)
Capacity: 414ci (6,458cc)
PTO power: 131hp @ 2,400rpm
Drawbar power: 113hp
Fuel consumption: 15.31hp/hr per gall
Weight: 12,715lb (5,722kg)

Diesel power came a long way in the 1950s. At the start of the decade, diesels were a costly option which not all manufacturers offered. By 1960, no mainstream tractor line-up was without a diesel version. A year later, Allis-Chalmers unveiled the D19 turbo diesel, offering gasoline power with diesel economy. Its rivals had no choice but to follow suit.

International's answer came four years later, in the shape of the 1206 – but this was no too little-too late attempt to keep up with Allis. The A-C D19 was a

66hp tractor, but the new International had over 100hp. It was the company's most powerful machine yet, and caused a real stir at a time when most tractors were in the 40-70hp class. It was developed in house, using a turbocharged version of International's existing D361 six-cylinder engine (already used in the 806 tractor). The transmission was beefed up to suit, with hardened gears, heavier pinions and final drive gears. Tyres had to be specially designed as well – the prototype 1206 had so much power that conventional tyres buckled their sidewalls or simply spun off the rim. A new tyre had to be designed specifically for the 1206.

The 1256 pictured here was an update, unveiled in 1967, and part of a whole range of 56-series tractors. Most significantly, the turbo diesel was boosted in size to 407ci (6,349cc), with a larger bore and stroke. A 'Deluxe' two-door cab was optional, as was two-post ROPS (roll-over protection) and air conditioning.

Below: Turbo'd 1256 needed special tyres to cope with its power.

International 784

USA

Specifications for International 784 petrol (1979)
Engine: Water-cooled, four-cylinder
Bore x stroke: 3.9 x 5.0in (98 x 125mm)
Capacity: 246ci (3,838cc)
PTO power: 65.5hp @ 2,400rpm
Drawbar power: 56.8hp
Transmission: Eight-speed
Speeds: 1.7-15.6mph (2.7-25.0km/h)
Fuel consumption: 15.34hp/hr per gallon
Weight: 6,090lb (2,741kg)

'You get more in a new 84' went the ad slogan, when International's 84-series was launched in 1978. This was a new range of small row-crop and utility tractors, designed to replace the 74-series and built at the IH factory in Doncaster, England.

There were six models, of which the 784 pictured here was the largest and most powerful. The range kicked off with the 36hp 384 Utility, powered by a 154ci four-cylinder diesel. Then came the 484, whose 179ci three-cylinder diesel offered 42hp, and came with a three-point hitch, adjustable wide front axle and 20-gallon fuel tank. Next up, the 52hp 584 and 62hp 684 used 206 and 239ci diesels respectively, both with eight forward and four reverse speeds. There was also a hydrostatic drive version, built to the same specfication as the 684 but with a hydrostatic transmission replacing the gearbox. Top of the range was the 784 shown here, which in diesel form used International's own 246ci diesel, offering 65hp – as you can read in the specifications, a gasoline version was still optional, though by now most farmers were opting for diesel power.

So popular was the 84-series that, styling apart, the basic designed kept rolling off the IH production lines right through the 1990s.

Below: 84-series underpinned the smaller Internationals for nearly 20 years.

International 1486

USA

Specifications for International 1466 turbo diesel (1973)
Engine: Water-cooled, six-cylinder, turbo
Bore x stroke: 4.3 x 5.0in (108 x 125mm)
Capacity: 436ci (6,802cc)
PTO power: 145.9hp @ 2,600rpm
Drawbar power: 123.2hp
Transmission: Sixteen-speed
Speeds: 1.5-22.3mph (2.4-35.7km/h)
Fuel consumption: 13.03hp/hr per gallon
Weight: 13,670lb (6,152kg)

If the 1960s were a decade of ever increasing power, the '70s saw new attention given to driver comfort. Since World War II, the industry had gradually become more aware of the need for ROPS (roll-over protection) but now a more affluent society saw no reason why tractor drivers should be jolted, jarred, or frozen more than anyone else!

This was the thinking behind International's Pro Ag line, the 56-series tractors unveiled in September 1976. Its cab was all new, still with two doors (one each side) but a larger glass area of 43.2 square feet. ROPS protection was built in, exceeding the SAE standards of the time. Significantly, the driver's seat was moved forward by eighteen inches compared to 66-series. Sitting forward of the rear axle made for a smoother ride, not to mention easier entry and exit. The doors were of double-wall construction, with heavy rubber seals on them to keep dust out. Whenever the doors were shut, an air filter self-cleaned the cab of any dust that had found its way in. Air conditioning was standard, and a radio, tape player and hydraulically mounted seat were all on the options list.

There were technical advances too, and the 86-series offered an optional (later standard) digital control centre. This gave readouts of engine rpm, PTO speed, exhaust gas temperature and ground speed. The 1486 shown here was powered by the 145hp DT414 turbo diesel.

Below: Not all 86-series Internationals had air conditioned, dust-free cab.

Ivel

UK

Specifications for Ivel 'Agricultural Motor'
Power: 20hp @ 850rpm
Weight: 3,638lb (1,654kg)

There are many claims as to who produced the first practical tractor. Hart-Parr and Allis-Chalmers both have good cases, but a forgotten Englishman named Daniel Albone may have pre-dated even them. Hailing from Biggleswade in Hertfordshire, Albone was a racing cyclist and inveterate inventor. He produced bicycles (including the first ever ladies' frame), tandems, motorcycles...and a tractor.

In 1902 he patented what he called the 'agricultural motor'. which he named the Ivel after a river near his home – the term 'tractor' hadn't yet been thought of! But this was no paper patent. The Ivel tractor actually existed, and went into production at Albone's factory the year after his patent was filed. With 20hp at 850rpm, the Ivel was a tricycle design which weighed 3,638lb (1,654kg). Sadly, its inventor died in 1906, and without his driving force his company went into receivership in 1920. But he left us with what was arguably the world's first practical tractor, powered by an Otto-cycle engine.

Below: Another claimant to the title of first-ever tractor, the 1903 Ivel.

JCB

UK

Specifications for JCB Fastrac 2150 (1998)
Engine: Water-cooled turbo diesel
Engine: 150hp
PTO power: 133hp
Torque: 467lb ft @ 1,400rpm
Transmission: 54 forward, 18 reverse speeds
Weight: 14,032lb (6,378kg)

JCB – JC Bamford – will forever be associated with the big yellow 'digger' that made it famous. But in more recent years it has branched into agriculture with the Fastrac range of (as the name suggests) sophisticated four-wheel tractors. The concept of the Fastrac is a fully capable agricultural machine which can travel on-road at up to 45mph, thus reducing hauling times and (as a side effect) frustrating fewer motorists. It bristles with technology. All Fastracs are

powered by the Perkins 1000 turbo diesel engine, some with intercoolers, and power ranging from 115-170bhp. The ROPS (rollover protective structure) cab is air conditioned, and unusually has a passenger seat. JCB's Quadtronic four-wheel steering system automatically switches between two- and four-wheel-drive for quicker turns at the end of the field. The turning circle is particularly tight, thanks to three-link front suspension, while the rear suspension is self-levelling to compensate for the additional weight of implements. The whole suspension system is even self-levelling from side to side, which is helpful when traversing hillsides. A radar slip control compares actual ground speed to engine speed, and there are external disc brakes. The rear axle has a soft engage differential lock, which can be switched in even while the wheels are spinning, without damaging the drive-train.

Below: Fastrac combines fast on-road performance with field work.

John Deere

USA

John Deere was a blacksmith, and the company he founded in 1837 is still going strong today – it's survived innumerable crises and setbacks, the only big US tractor maker still in business under the same name. As the southern prairies were opened up to the plough, John Deere set up a thriving business making steel plough shares. But his didn't pioneer tractors – not until 1914 was board member John Dain given the task of building one. The problem was the three-wheel Dain-Deere retailed at $1,700 when a Fordson cost less than $700. The answer lay in the Waterloo Gasoline Engine Company, already producing a

practical 25hp machine. John Deere bought up the whole thing and finally entered the tractor business proper. So successful was the Waterloo's twin-cylinder format that John Deere stuck with it right up to 1960. Since then, it has developed ever more sophisticated three-, four- and six-cylinder machines, from its plants in the US, Germany, Spain, South Africa, Argentinia, Mexico and Australia. An American company with a worldwide base.

Below: Few colours are more instantly recognisable than JD green and yellow.

John Deere Waterloo Boy

USA

Engine Type: Water-cooled, twin-cylinder ohv
Bore x stroke: 5.5 x 7 inches (138 x 175mm)
Capacity: 333ci (5,199cc)
Fuel: Kerosene
PTO power: 25hp @ 750rpm
Drawbar power: 12hp @ 750rpm
Transmission: Two-speed

The Waterloo Boy, as it was named, was unlike many other tractors, with its twin-cylinder overhead valve engine mounted horizontally in the frame. This layout would actually be Waterloo's legacy to JD right up to the early 1960s.

After several design changes, the Model R of 1914 ended up with a 333ci (5.2 litres) version of the twin-cylinder engine, with 25 belt hp and 12 drawbar hp at 750rpm. There were actually thirteen variations on the Model R theme before the N Model took over in 1917. Its twin-cylinder engine was to grow in both size and power. It also added a two-speed transmission (the R was a single-speed) which led to an easy way to recognise the difference between the two – the N has a huge drive gear for each rear wheel, nearly as big as the wheel itself.

Compared to some of its smaller, lighter rivals, the Waterloo Boy looked a little clumsy and old-fashioned. But for John Deere, its existence was crucial – it gave the company a toehold in the tractor market and led directly to the twin-cylinder format that underpinned JD machines for the next 40 years.

John Deere GP WT

USA

Engine: Water-cooled twin-cylinder, side-valve
Bore x stroke: 5.75 x 6in (144 x 150mm)
Capacity: 311ci (4,852cc)
Fuel: Kerosene
PTO power: 20hp @ 950rpm
Drawbar power: 10hp @ 950rpm
Transmission: Three-speed

The WT was John Deere's Wide Tread version of the standard GP. It came about because farmers complained that the standard machine's broad bonnet and radiator coupled with its standard front axle hampered their forward visibility when cultivating. This was particularly so in the south of the USA, where cotton growers preferred the Farmall, with its tricycle front end and wide track rear that came in two- or four-row form – the GP was a three-row tractor only.

John Deere's answer gave the wide tread and two-wheel tricycle front end of the Farmall, but on the established GP chassis. Like the GP, the WT used the 311ci (4.9 litres) twin-cylinder motor, upsizing to a 339ci (5.3 litres) in 1931. Another big change was to an overhead steering linkage in 1932. The original side-mounted system, with the linkage running down the offside of the machine, was subject to front-wheel whip.

Heavier than the standard GP, the WT was geared for slightly lower speeds, though it retained the three-speed transmission and chain final drive. More noticeably, it was over 26in wider than the original, with an 8-inch longer wheelbase, and the rear tread width of 76in allowed it to straddle two rows, just

Above: Twin-cylinder Waterloo Boy was JD's path into the tractor business.

Above: Narrower front end allowed better crop visibility for the Wide Tread.

like the Farmall. To go with the WT, John Deere introduced a new series of two- and four-row implements. The GP400, for example, was a four-row cotton and corn planter which allowed one farmer to 'plant from 35 to 45 acres per day.'

Unfortuately for John Deere, the WT failed to topple the Farmall from its wide-tread pedestal, but it did regain some of the ground lost with the first GP. If nothing else, it proved that the company was responsive to criticism, and many of its features were carried over to the replacement Model A.

John Deere GP

USA

Engine: Water-cooled twin-cylinder, side-valve
Bore x stroke: 5.75 x 6in (144 x 150mm)
Capacity: 311ci (4,852cc)
Fuel: Kerosene
PTO power: 20hp @ 950rpm
Drawbar power: 10hp @ 950rpm
Transmission: Three-speed
Speeds: 2.3-4.3mph (3.7-6.9km/h)
Weight: 3,600lb (1,636kg)

There aren't many names more famous among tractors than the John Deere GP. It was Deere's first successful attempt to produce a General Purpose (hence GP) row-crop machine. Originally introduced as the Model C in 1927, it was soon renamed GP because (so the story goes) there was confusion between 'C' and 'D' when dealers put in their orders over the phone!

There were five versions of the GP, all of which used Deere's famous twin-cylinder format. Long after rivals had gone over to four- or even six-cylinder motors, JD stuck with the twin – not as smooth or powerful as younger rivals,

it was at least reliable, easy to service and economical on fuel. However, history could easily have been different – the only reason JD chose the twin in the first place was because of the large inventory of engines left over from the unsuccessful Waterloo Boy – it couldn't afford to build a new one!

In the first GP, it was a side-valve engine of 311ci (4.9 litres), designed to run on kerosene but needing higher octane gasoline start up. Pre-ignition was a common problem with the low grades of fuel, so the GP was fitted with a water valve by which the driver could send a squirt of water directly into the combustion chamber, damping down the pre-ignition. At 950rpm under load, the first GP produced 20 belt hp, 10 drawbar hp, though a more powerful version arrived in 1931. A power take-off was optional, but the GP's most innovative feature was its mechanical power lift. By means of a foot pedal, attachments could be raised and lowered, using engine power – an option at first, it soon became standard on nearly all general purpose tractors. The GP was certainly a step forward for John Deere, but compared to the Farmall it lacked power and ground clearance, reflected in its sales.

Below: Twin-cylinder GP used a mechanical implement lift.

John Deere Model D (unstyled)

USA

Specifications for John Deere D (1940)
Engine: Water-cooled, twin-cylinder
Bore x stroke: 6.8 x 7.0in (170 x 175mm)
Capacity: 508ci (7,925cc)
PTO power: 40hp @ 900rpm
Drawbar power: 34.5hp @ 900rpm
Transmission: Three-speed
Speeds: 3.0-5.3mph (4.8-8.5km/h)
Fuel consumption: 10.1hp/hr per gallon
Weight: 8,125lb (3,693kg)

If ever a tractor with a John Deere badge deserved the epithet 'milestone', it's

this, the D model. Not only was it the first machine to bear the John Deere name (apart from the unsuccessful four-cylinder Dain-John Deere) but it established the standard JD layout which was to form the company's backbone for the next 40 years. And with a 30-year production run, the D was John Deere's longest running tractor as well. It was a hit.

The secret, which ensured JD's success over the next few decades, was that of a basic, simple machine that placed rugged reliability above everything else. The twin-cylinder motor (when most rivals were using fours) was mounted horizontally in the frame; it had an external flywheel and was slower-revving than any rival. There were just two transmission speeds; and in the manner of early tractors, the D ran on kerosene (cruder but cheaper than gasoline) but had to be started on gasoline. Switch over to kerosene, said the handbook, when the radiator was 'too hot to hold your hand on.'

Below: Twin-cylinder Model D established the classic John Deere layout.

John Deere A Model

USA

Engine: Water-cooled twin-cylinder, ohv
Bore x stroke: 5.5 x 6.5in (138 x 163mm)
Capacity: 206ci (3,214cc)
Fuel: Gasoline or kerosene
PTO power: 24hp @975rpm
Drawbar power: 16hp @ 975rpm
Transmission: Four-speed (later Six-speed)
Speeds: 2.0-6.3mph (3.2-10.1km/h)
Weight: 4,088lb (1,858kg)

The A Model nearly never happened at all. In the early 1930s, the Depression devastated the US tractor industry – production slumped to a low of around 20,000 (one-tenth of what it had been a couple of years earlier) and John Deere, like almost everyone else, was losing money fast. But the board was persuaded to carry on investing in two new models: the A and the B.

Launched in 1934, the A's most innovative features included an adjustable rear axle on most variants), centre-line hitch and PTO, and the first fully hydraulic power lift system. Some things didn't change – it used the

traditional Deere twin-cylinder engine (simpler than the four-cylinder opposition) albeit in updated overhead valve form. But it was still a simple engine, using thermo-syphon cooling instead of a water pump. At first, it was rated at 24 belt hp, 16hp at the drawbar, but a bigger version in 1939 (using a longer stroke and bigger valves) recorded 26hp/20hp. Later still (it was 1947) a higher compression of 5.6:1 increased power again, to 33 belt hp and 26hp at the drawbar. That engine was matched to a six-speed transmission, which replaced the earlier four-speed.

There was of course a whole range of different A Model variants, but the AO pictured here was designed for grove, orchard and vineyard work. Introduced in 1935, it lacked the adjustable rear axle seen on many Model As, but the AO Streamlined offered a rounded front grille, bonnet and rear mudguards. Along with the standard AO, it was a low-profile machine with lowered exhaust, air intake, steering and controls, all designed not to snag on branches. The AO was restyled in 1949 and replaced by the Model 60 four years later.

Below: Despite the Depression, JD persevered with the overhead valve A.

John Deere B Model

USA

Engine: Water-cooled twin-cylinder, side-valve
Bore x stroke: 4.25 x 5.25in (106 x 131mm)
Capacity: 149ci (2,324cc)
Fuel: Gasoline or kerosene
PTO power: 14hp @ 1,150rpm
Drawbar power: 9.4hp @ 1,150rpm
Transmission: Four-speed (later Six-speed)
Speeds: 2.3mph-5.0mph (3.7-8.0km/h)
Weight: 2,731lb (1,241kg)

John Deere's B was basically a smaller version of the Model A, about 'two thirds the size in power and weight', according to the publicity, but with al the A's advanced features.

Smaller tractors usually mean smaller profits, but in America it made sense to build one. In 1930, although small farms of less then 100 acres made up only 15.7 per cent of the land under tillage, they accounted for over half of the farms. So small farmers were a significant part of the tracto market – what they wanted was a single-plough general purpose machine

and that's what the Deere B gave them.

But despite its small size, the B was a genuine miniature of the bigger John Deeres, with a little 149ci (2.3 litres) version of the classic twin-cylinder water-cooled engine, though unlike the A, it retained the side-valve format of older JDs. It ran slightly faster than its bigger brothers – 1,150rpm – and produced 14hp at the belt, 9.4hp at the drawbar, according to Nebraska.

As with the A, there were many variations on the theme: the basic B used a two-wheel tricycle front end; BN, single front wheel; BW, wide front model; BR, standard front; BO, orchard version; BNH, high crop version of the BN; BWH, high crop version of the BW; BO Lindemann, a crawler model converted by Lindeman Power & Equipment; finally, BI, the industrial version. They all benefited from a larger engines: 175ci (2.7 litres) in 1939 and 190ci (3.0 litres) in 1947 – the latter was offered in both all-fuel and high compression gasoline versions. As with the A, a six-speed transmission later replaced the four.

Below: The B was a scaled-down Model A, with most of the same features.

John Deere Model D (styled)

USA

Engine: Water-cooled, twin-cylinder
Bore x stroke: 6.8 x 7.0in (170 x 175mm)
Capacity: 508ci (7,925cc)
PTO power: 40hp @ 900rpm
Drawbar power: 34.5hp @ 900rpm
Transmission: Three-speed
Speeds: 3.0-5.3mph (4.8-8.5km/h)
Fuel consumption: 10.1hp/hr per gallon
Weight: 8,125lb (3,693kg)

So the John Deere Model D was a huge success – over 1,000 were sold in the first year and by the time production ceased in 1953, over 160,000 had been built. The D came to be seen as one of the definitive ploughing tractors, or a do-it-all for threshing.

Of course, there were plenty of changes over the years. Many small engine improvements saw power rise from 27hp at first to 42bhp by 1953. A third transmission speed was added in 1935, and this styled version joined the range four years later. The addition of extra bodywork, by well known industrial designer Henry Dreyfuss, had a dramatic effect on the looks of the D – although the basic layout stretched back to 1923 and beyond, the bodywork D of 1939 looked right up to date, especially with the optional pneumatic rubber tyres. The story goes that one of JD's engineers turned up at Dreyfuss' New York office in a fur coat and straw hat. This so impressed the designer with the potential for redesigning a tractor that he agreed to the job right away!

This was the tractor that established John Deere as a mainstream mass maker of tractors. It helped the company survive against tough competition from Fordson, JI Case and Farmall – a true landmark tractor.

Left: Industrial designer Henry Dreyfuss restyled the D in 1939.

John Deere G Model

USA

Engine: Water cooled twin-cylinder, ohv
Bore x stroke: 6.125in x 7in (153 x 175mm)
Capacity: 410ci (6,396cc)
PTO power: 31hp @ 975rpm
Drawbar power: 21hp @ 975rpm
Transmission: Four-speed (Six-speed from 1942)
Speeds (4spd): 2.3mph-6mph (3.7-9.6km/h)
Weight: 4,400lb (2,000kg)

In its time, the John Deere G was the largest row-crop tractor you could buy. Launched in 1937, it reflected a general move towards larger farms on which fewer people worked.

It certainly had plenty of power to tackle big acreages, notching up 31hp at the PTO on its first Nebraska test and 21 drawbar hp. The source was John Deere's long-familiar horizontal twin-cylinder engine measuring 410ci, its biggest form yet. It was lower revving than equivalent four-cylinder motors, rated at 975rpm, but wasn't lacking in power or torque, being the first over-30hp general purpose machine. While other tractor makers were following Oliver into the new streamlined era, the G remained unstyled until 1942. But it wasn't behind the times. The original four-speed gearbox was replaced with a six-speed that same year, and rubber tyres were standardised. Final drive was by spur gears.

John Deere BWH

USA

Engine: Water-cooled twin-cylinder, side-valve
Bore x stroke: 4.25 x 5.25in (106 x 131mm)
Capacity: 149ci (2,324cc)
Fuel: Gasoline or kerosene
PTO power: 14hp @ 1,150rpm
Drawbar power: 9.4hp @ 1,150rpm
Transmission: Four-speed (later Six-speed)
Speeds: 2.3mph-5.0mph (3.7-8.0km/h)

The BWH and sister BNH were no more nor less than high-crop versions of the standard BN and BW. These in turn were variations of the standard Model B, with narrow and wide-front axles respectively. The BW (offered all the way through from early 1935 to the summer of 1952) front end could be adjusted between 56 and 80 inches (1,400-2,000mm), in four-inch steps – it was a useful feature, as setting the front axle to the same width as the rear reduced soil packing.

The BWH high-cropper didn't arrive until five years later, and together with the BNH was a hybrid of parts taken from the B, BN and BW, with some all-new components as well. Engine and transmission-wise, it was the same as the standard model B. The result was an extra two inches of clearance under the rear axle, a result of using the same 40-inch drive wheels as the BNH, in place of the standard 36-inch ones. And there was an extra three inches of clearance at the front, thanks to lengthened front axle knuckles.

The front tread was slightly bigger as well, being adjustable between 42.6 inches and 54.6 inches(1,065-1,367mm) (BW, 40-52in, 1,000-1,300mm)). If that wasn't enough, 7 and 13-inch extension assemblies were available, giving front tread adjustments of 56.6-68.6in (1,415-1,715mm) and 68.6-80.6in (1,715-

Above: At 6.3 litres, the G was John Deere's biggest tractor yet.

The rear wheel tread was adjustable between 60 and 84 inches (1,500-2,100mm). That was on the standard axle – anything up to 112 inches was possible with optional axles. A power take-off was standard, and hydraulics optional. There were some overheating problems early in the Gs life, but a larger capacity radiator solved that, after about 3,000 Gs had left the line. Another 60,000 Gs would be built before production ended in 1953.

Above: The BWH denoted wide front axle and high-clearance.

2,015mm) respectively. All versions of the BWH were built on a longer chassis than the standard machine. There were also a small number of BWH-40s built, with rear wheel tread adjustable down to a narrow 42.5 inches. Mudguards were standard, as on that narrowest setting the drive wheels were very close to the operator.

John Deere Model R Diesel USA

Engine Type: Water-cooled twin-cylinder diesel
Capacity: 416ci (6,490cc)
PTO power: 51hp
Drawbar power: 46hp

If the D was one of John Deere's landmark machines, then the R was another, the biggest machine JD had ever built, and its first ever five-plough tractor. But there was something else even more significant – it was a diesel.

Even before World War II, there were signs of a horsepower war among tractor makers, just as there had been among cars. The reason was simple – by the late 1930s, American farms were getting larger, and bigger fields justified bigger tractors that could pull bigger ploughs, pull them faster and not get bogged down. After the war, this power rivalry proceeded apace, but it was leading to ever higher fuel consumption

from the gasoline and distillate engines.

John Deere already had an answer on the drawing board during the war years. The radical new R brought with it a diesel version of the classic Deere twin-cylinder motor. At 416ci (6.5 litres), it was bigger than any previous JD, and more powerful as well, at 51PTO hp. But it was more fuel-efficient than any other big tractor. There were other innovations as well: it had a live power take-off and a live Powr-Trol hydraulic system. It was as if, after two decades of producing the simplest tractors, John Deere was determined to head right up to date. Better still, the R was cheaper to run than any comparable rival. Farmers flocked to buy them and over 20,000 Rs were built between 1949 and 1951. It was some years before diesel became universal among tractors, but the John Deere R started it all.

Below: Dieselising the JD twin-cylinder engine brought more power.

John Deere 80 Diesel

USA

Engine: Water-cooled twin-cylinder diesel
Bore x stroke: 6.125 x 8in (153 x 200mm)
Capacity: 475ci (7,410cc)
PTO power: 65hp @ 1,125rpm
Drawbar power: 60hp @ 1,125rpm
Transmission: Six-speed
Speeds: 2.5mph-12.5mph (4.0-20km/h)
Fuel consumption: 15.96hp hr/gal
Weight: 8,511lb (3,869kg)

The early 1950s were busy times for John Deere, when its entire range was updated. The R model, being relatively new, was the last in line for this treatment, but in 1955 was replaced by the 80. Just like the R, this was powered by the biggest John Deere twin, with each cylinder now measuring over six inches across, running up and down a bore of eight inches. To the casual observer, such massive

cylinders might look like a desperate attempt to squeeze competitive power from an outmoded design. In reality, the 80 Diesel set a record for lowest fuel consumption on a lb/hp basis when tested at Nebraska.

In truth, the 80 had much in common with the machine it replaced. There was a new six-speed transmission, and the R's ground-breaking twin-cylinder diesel engine was increased in size to 475ci (7.4 litres). But otherwise it was business as usual. One interesting point was that although the R had been equipped with an electric start, the 80's massive cylinders were simply too big for such a system. Instead, it had a small donkey engine – a 60ci (936cc) V4 petrol, which ran up to 5,500rpm just on starting duties! It was actually the Waterloo factory's first shot at building an engine with more than two cylinders, and it worked. The little V4's exhaust was routed so that it would help warm up the big diesel before attempting to start it. Still, it was a mechanically complex way of starting up each morning, and a heavy-duty 24-volt electric starting system later superseded the V4.

Below: John Deere's biggest twin was this 7.4 litre diesel fitted to the 80 of 1955. Starting was via a small V4 donkey engine, and later 24-volt electrics.

John Deere 4320

USA

Specifications for John Deere 4320 Syncro-Range diesel
Engine: Water-cooled, six-cylinder
Bore x stroke: 4.25 x 4.75in (106 x 119mm)
Capacity: 404ci (6,302cc)
PTO power: 108hp @ 2,200rpm
Drawbar power: 102hp @ 2,200rpm
Transmission: Eight-speed
Speeds: 2.0-18.9mph (3.2-30km/h)
Fuel consumption: 15.4hp/hr per gallon
Weight: 10,675lb (4,852kg)

In 1960, the John Deere range was renewed from head to toe. Out went the old twin-cylinder machines, in came an all-new line-up of four- and six-cylinder tractors. The product of a seven-year development programme, they ranged from 35 to 80hp and came with new styling to underline what a transformation this was. Ninety-five per cent of the parts were new. As part of the development process, John Deere had asked farmers what they wanted from a tractor: more power, bigger fuel tanks, better transmissions and hydraulics, and extra comfort for the driver, came the answers. With the new 10 series, they got what they asked for.

But only three years later, it was replaced by the 20-series. The four-cylinder 3020 and six-cylinder 4020 had Powershift shift-on-the-go transmission (with eight forward speeds), soon joined by a more powerful 5020. The 4020 in particular was a great success, accounting for nearly half of JD's entire sales in

John Deere 420

USA

Specifications for John Deere 420 (1956)
Engine: Water-cooled, twin-cylinder
Bore x stroke: 4.25in x 4.0in (106 x 100mm)
Capacity: 113ci (1,762cc)
PTO power: 29.2hp @ 1,850rpm
Drawbar power: 27hp @ 1,850rpm
Transmission: Four-speed (Five-speed option)
Weight: 3,250lb (1,477kg)

When the John Deere 420 was introduced in 1956, the decision to replace the long-running twin-cylinder line had already been taken. But the revolutionary new range was still four years away, and in the meantime JD needed to breathe a little more life into the older models, to see them through the last few seasons.

In fact, the whole range was restyled and renumbered as the 20-series, replacing the previous two-numbered machines which had been on sale for three years. All of these were direct replacements or updates of existing tractors. So the 420 replaced the Model 40, and the 520, 620 and 720 took over from the 50, 60 and 70 respectively. A new smaller general purpose tractor – the 320 – was added at the same time.

Spotting the 20-series is easy – they were the first John Deeres to use that yellow stripe on the bonnet. But technically, the biggest news was the Custom Powr-Trol, JD's version of the Ferguson Draft Control. Harry Ferguson had settled his agreement with Ford, which meant the end of patent restrictions – so just like Draft Control, the Powr-Trol automatically raised the implement a

Right: Six-cylinder 4320 was a far cry from the early twins.

the US and Canada in 1966. Also from that year it had the option of power front-wheel-drive, offering a 20 per cent boost to traction.

But the 1970s saw demands for yet more power, and the 4320 pictured here was uprated to 115hp, up from the standard 94hp. Alongside it was the 4620, with a turbo-intercooled version of the same engine giving 135hp.

Above: This was the penultimate John Deere twin, the 20-series.

little when it hit heavy ground, to maintain speed. Once the going was easier, the draft returned back to its pre-set level. Meanwhile, the engine might have two cylinders, but they were bigger, and new pistons and heads meant more power and lower fuel consumption.

John Deere 4430

USA

Specifications for 4430 diesel (1972)
Engine: Water-cooled, six-cylinder, turbo
Bore x stroke: 4.25 x 4.75in (106 x 119mm)
Capacity: 404ci (6,302cc)
PTO power: 126hp @ 2,200rpm
Drawbar power: 105hp @ 2,200rpm
Transmission: Sixteen-speed (eight-speed standard)
Speeds: 2.0-17.8mph (3.2-28km/h)
Fuel consumption: 15.6hp/hr per gallon
Weight: 11,350lb (5,159kg)

'New Generation' had been the name given to the four- and six-cylinder tractors that transformed John Deere's range overnight in 1960. Twelve years later, the 'Generation II' 30-series arrived. The 10- and 20-series hadn't fallen behind in mechanical terms – power and transmission options were often updated through the 1960s. But now there were higher standards of driver comfort to live up to, quite apart from a growing demand for roll-over protection.

JD's response was the Sound-Gard cab. With its large curved windscreen and tinted glass, this looked luxurious, and it was quiet enough to bring the option of a stereo radio-cassette. It also had built-in roll-over protection, and the option of a pressurizer to keep dust and dirt out, or air conditioning. Underslung pedals, seat belt and adjustable steering wheel were part of the 30-series package and over half of buyers opted for the Sound-Gard cab as well. It was a real step forward in making life safer, quieter and more comfortable for the driver.

Of the new 30-series, the 4430 shown here was the most popular. It used the turbocharged version of JD's six-cylinder diesel, to give 125hp. If that wasn't enough, the intercooled 4630 offered 150hp. Also new was Perma-Clutch, with oil cooling to maximise clutch plate life, and a sixteen-speed Quad Range transmission, which added a two-range clutchless change to the existing eight-speed Syncro-Range.

Below: Sound-Gard cab from 1972 had built-in roll-over protection.

John Deere 830 Diesel

USA

Engine: Water-cooled,three-cylinder
Bore x stroke: 3.86 x 4.33in (96.5 x 108mm)
Capacity: 152ci (2,371cc)
PTO power: 35hp @ 2,400rpm
Drawbar power: 28hp @ 2,400rpm
Transmission: Eight-speed
Speeds: 1.9-15.9mph (3.0-25.4km/h)
Fuel consumption: 14.3hp/hr per gallon
Weight: 4,376lb (1,989kg)

John Deere had taken over Lanz of Germany in 1959, and from then on the Lanz factory at Mannheim gave JD a solid European manufacturing base. From here, tractors were made to suit European conditions, though there was plenty of crossover – some Mannheim tractors used American-made engines, for example, while some US tractors were exported to Europe.

The 830 was part of the 30-series unveiled by Mannheim in 1972, to

replace the 20-series which had arrived five years earlier. Neither of these of course, should be confused with the American made 20 and 30 series of the mid-1950s – these were the last of the old twin-cylinder tractors, completely different from the Euro machines. As announced, the 20-series made a complete range, from the three-cylinder 34PS 820, to the 64PS four-cylinder 2020. Bigger machines – the 3020, 4020 and 5020 – were all imported from the American Waterloo factory. Mannheim's first six-cylinder tractor,the 2120, was added to the range a year later.

The 30-series of 1972 was mechanically similar to these, with the 830 the smallest of the range, featuring two-wheel-drive and four-post ROPS rather than a full cab. All other 30-series had mechanical four-wheel-drive, apart from the top model 2130 and 3130, which featured hydrostatic front-wheel-drive.

Below: Final US-built 830. Not to be confused with the later German-built 830.

John Deere 8850

USA

Specifications for 8850 Diesel (1982)
Engine: Water-cooled V8, turbo intercooled
Bore x stroke: 5.5 x 5.0in (138 x 125mm)
Capacity: 955ci (14,898cc)
PTO power: 304hp @ 2,100rpm
Drawbar power: 270hp @ 2,100rpm
Transmission: Sixteen-speed
Speeds: 2.1-20.2mph (3.4-32.3km/h)
Fuel consumption: 14hp/hr per gallon
Weight: 37,700lb (17,136kg)

In the new breed of super-tractor that developed through the 1970s and '80s, power once again became the watchword. It often seems strange to the outsider that, with farming in economic turmoil for much of this time, tractor makers should carry on developing ever bigger, more expensive and complex machines. But on big farms, bigger tractors could do more work, faster and cheaper than a smaller equivalent – they paid for themselves, which was the

bottom line. The quest for power became such that by the late '70s even six turbocharged cylinders sometimes weren't enough.

In 1982, John Deere responded with a new top-of-the-range V8 super-tractor, the 8850, which ran alongside the existing six-cylinder 8450 and 8650. Instead of buying in a V8 diesel from outside, JD designed its own, a turbocharged and intercooled diesel with 300hp at the PTO. All three of these four-wheel-drive machines were given improved visibility (exhaust pipe and air cleaner intake were moved to the right-hand side) and new ISO remote hydraulic couplers. A 16-speed Quad-Range transmission was standard, controlled by a single lever, and just to prove that fuel economy was an issue as well as power, the 8850 had a viscous fan drive that reduced fan speed in cooler temperatures. And just in case bystanders mistook the 8850 for its 'cheaper' brothers, it was given six headlights in place of the their four – who says farmers aren't as status-conscious as the rest of us?

Below: John Deere designed, developed and built its own V8 diesel for the 8850.

John Deere 6400

USA

Specifications for 6420 (2002 model)
Engine: Water-cooled, four-cylinder, turbo
Bore x stroke: 4.2 x 5.1in (106 x 127mm)
Capactiy: 282ci (4,530cc)
Power: 120hp @ 2,200rpm
Torque: 337lb ft @ 1,495rpm
Transmission: Twenty-four-speed
Speeds: 1.0mph-25mph
Weight: 10,350lb (4,750kg)

The 1980s were not kind to American tractor manufacturers. With falling prices and stiff competition from imports, farmers (if they survived at all) had to cut back – some tractor dealers gave up and many manufacturers found themselves merged, taken over or just closed down. Through all of this, John Deere managed not just to survive, but keep its independence. The worldwide manufacturing and marketing base helped, but it also came down to regular updating of the range. Indeed, back in the plough-making days, Mr John Deere had said that the best way to stay in business was to continually improve the

product – so, more than a century later, that's what they did.

The new 50 series announced in the early '80s gave a slight power boost across the range – 5-10hp – and a new option of mechanical front-wheel-drive, which gave a tighter turning circle than the hydrostatic type. Sound Gard cabs were optional, but all the 50 series had on-the-go shifting. This in turn was replaced by the 55 series in 1987-9, again with a power boost across the range, an electronic monitoring system and automatic engagement of front-wheel-drive.

Nineteen-ninety-two was a busy year. The three biggest machines were updated as the 60 series and there was an all-new small/mid-range tractor built in Atlanta, Georgia. The latter was big news. During the 1980s, it had become accepted wisdom that it was no longer economically viable to build smaller tractors in the US, yet the 5000 series was a complete range of three-cylinder machines, all with power steering, nine-speed transmissions, and all made in the USA. Soon after the 5000 was launched, John Deere surprised the world with another new range – the 6000/7000 series, of which the 100hp Mannheim-built 6400 is shown here.

Below: Regular updates kept JD competitive through the tough 1980s.

John Deere 6900

USA

Specifications for 6920 (2002 model)
Engine: Water-cooled, six-cylinder, turbo
Bore x stroke: 4.2 x 5.1in (106 x 127mm)
Capactiy: 435ci (6,788cc)
Power: 150hp @ 2,100rpm
Torque: 499lb ft @ 1,365rpm
Transmission: Twenty-speed
Speeds: 1.6mph-31mph (2.6-50km/h)
Weight: 12,936lb 5,880kg

The all-new 6000/7000 series of 1993 was, 'the most revolutionary development in John Deere tractor models since the New Generation multi-cylinder models replaced the two-cylinder line in 1960."' So said Don MacMillan, the English farmer, John Deere dealer, user and author.

Actually, they weren't 100 per cent all-new – a whole nine components

were carried over from the 60-series. The 6000/7000 series were mid/high-range machines, the four-cylinder 6000s built in Mannheim and six-cylinder 7000s at JD's Waterloo plant. The engines were certainly all-new, made in-house by Deere, and at launch four variations powered the 6000 range: a 75hp 6100, 84hp 6200, 90hp 6300 and 100hp 6400, the latter three all turbocharged.

It wasn't long before a six-cylinder 6000 became available though, and the 6900 pictured here is one such. It's a 1996 model, at the time the top of the range 6000, with a 435ci (6.8 litres) turbocharged engine producing 130 PTO hp at 2,100rpm. There was also the 81hp 6600 and 120hp 6800, both with turbos. The transmission was what Deere called PowrQuad: four ranges (changeable under load, without declutching) giving 20 speeds, plus the oil-cooled Perma-Clutch II. The 6000 series had effectively extended upmarket to meet the 7000.

Below: Six-cylinder turbocharged diesel made this the top-range 6000.

John Deere 7800

USA/GERMANY

Specifications for 7800 (1993)
Engine: Water-cooled, six-cylinder, turbo
Capacity: 487ci (7,636cc)
Power: 170hp
Transmission: Nineteen-speed

If the 6000s were the European end of this latest generation of John Deeres, the 7000s represented America. All had six-cylinder turbocharged diesels, and all were built at the Waterloo factory. There were three: the 110hp 435ci (6.8 litres) 7600, 125hp 487ci (7.6 litres) 7700 and 140hp 487ci (7.6 litres) 7800 – a range-topping 7800 is shown here. It had four-wheel-drive and a 19-speed Powershift transmission, plus dual-stage braking and wet discs.

But it wasn't just the engines that were new about the 6000/7000 series.

Remember the Sound Gard, which had set new cab standards back in 1972? John Deere hoped that this latest generation cab would do the same 21 years later. The ComfortGard cab was claimed, at 72dB, to be the quietest in the world. It had 40 per cent more space than the Sound Gard, and 29 per cent more glass area. The seat was air-cushioned and the instrument panel tilted with the steering wheel, so whatever shape or size you were, the view was the same. Transmission, hydraulics, PTO and electro-hydraulic hitch were all new as well.
Nearly a decade later, the 6000/7000 still form the backbone of John Deere's tractor range, from the 80hp 6120 to the 175hp 7810.

Below: John Deere's Waterloo factory was still producing tractors in the late 20th century. This 170hp 7800 was the top of the 1993 range.

John Deere 8100

USA

Specifications for 8120 (2002 model)
Engine: Water-cooled, six-cylinder, turbo
Bore x stroke: 4.6 x 5.2in (116 x 129mm)
Capactiy: 8,100cc
Power: 200hp @ 2,200rpm
Transmission: Sixteen-speed
Speeds: 1.2mph-26mph (1.9-41.6km/h)
Weight: 19,800lb (9,000kg)

Not to be confused with the giant four-wheel-drive super-tractors, this 8000 series were big row-crop machines, announced in 1995. All four versions used the same six-cylinder diesel in different states of tune, and the least expensive 8100, with 160hp, is shown here. There were also a 180hp 8200, 200hp 8300 and 225hp 8400 – the latter was the first 225hp row-crop tractor on the market.

A new 16-speed powershift transmission was standard across the range, and front-wheel-assist was an option, though standard on the 8400. A

rubber track option was added in 1997. Caterpillar had shown how rubber tracks could be useful on farm tractors, and John Deere's response underlined how much buyers like the option. Automatic transmission was to join the options list as well, just as it did for the 6000/7000 series, though of a less sophisticated type. To avoid confusion, the articulated four-wheel-drive machines were renamed the 9000 series, culminating in the 425hp 9400T.

In late 2001, John Deere unveiled its range for the new year, revealing another round of upgrades and new options. The 8000 series all enjoyed slight power increases, and a new 8520 was added, extending the power range up to 295hp. A new suspension option – Independent Link Suspension – was offered for the front axle, said to eliminate power 'hop', and among the more 21st century features was the Starfire positioning system, which used GPS to pinpoint a tractor's location in the field to within 50mm. Times have changed.

Below: The 8000 series set new power standards for row-crop tractors.

Lamborghini

ITALY

Specifications for Victory Plus 260 (2002)
Engine: Water-cooled, six-cylinder, turbo-intercooled
Capacity: 458ci (7,146cc)
Power: 260hp
Transmission: Forty-speed
Speeds: Up to 25mph (40km/h)
Weight: 18,590lb (8,450kg)

If it hadn't been for a dismissive remark by Enzo Ferrari (so the story goes) Ferruccio Lamborghini would never have built luxury sports cars. Lamborghini apparently complained that his Ferrari had a noisy gearbox. In that case, declared its maker, you should stick to tractors!

Whatever the truth of that story, we do know that Snr Lamborghini was in the tractor business long before he turned to cars. Born in 1916, Lamborghini was actually a farmer, though he studied mechanical engineering at Bologna and found his metier after World War II, converting military surplus hardware

into agricultural machinery – swords into ploughshares. In 1949, the Carioca model used a Morris engine, converted from petrol to diesel. But the following year saw the first all-Lamborghini tractor, the L33, though this too used a Morris-based engine, with the addition of Lamborghini's patented hot bulb. His company also began making its own transmissions that year – later (in 1966) Lamborghini tractors were the first in Italy equipped with a synchromesh gearbox.

Production was boosted to 1,500 tractors a year, but an economic slump brought a takeover by rival tractor maker SAME in 1972 – Lamborghini himself retired the following year. Under new ownership, Lamborghini tractors flourished – 10,000 were built in 1980, and a new range was launched in 1983. Four-wheel-drive, electronic shift control and other high-tech features were adopted by Lamborghini in the 1980s and '90s. The company is now part of the SAME/Deutz-Fahr group.

Below: Under SAME ownership, Lamborghini tractors flourished.

Landini 65F

ITALY

Specifications for Globus 65 T (2002)
Engine: Water-cooled, four-cylinder diesel
Capacity: 160ci (3,990cc)
Power: 67hp
Transmission: Twenty-five-speed (F & R)
Wheelbase: 112in (2,800mm)
Weight: 6,710lb (3,050kg)

Giovanni Landini was a blacksmith's apprentice who went into business on his own account in 1884, though he didn't begin working on a tractor until the early 1920s. Sadly, he never saw his last project through, as he died in 1924, but his sons persevered, and announced a 30hp semi-diesel machine the following year.

As a company, Landini was firmly wedded to the semi-diesel concept, which was popular with many European manufacturers in the 1920s,'30s and in some cases right through the '50s. Simpler and cheaper to make than a true diesel, they relied on a hot spot to ignite the fuel, rather than engine compression alone. They were smokey, rough and woefully inefficient, but cheap and simple.

Landini established itself with semi-diesels like the Velite, Buffalo and Super, but by the 1950s, the whole concept was looking outmoded next to modern multi-cylinder diesel tractors. A Dr Flavio Fadda took over Landini just in time, and negotiated a licencing deal with Perkins of England. From 1957, nearly all Landinis were powered by licence-built Perkins, and a range of modern tractors was designed to suit. The 65F pictured here is a good example of Landini's range of small machines, popular on the home market.

Right: Landini nearly collapsed in the mid-1950s, but a new range of modern tractors saved the company.

landini
advantage 65F

Landini 130 Legend

ITALY

Specifications for Legend 130 (2002)
Engine: Water-cooled, six-cylinder turbo-diesel
Capacity: 385ci (6.0 litres)
Power: 126hp @ 2,200rpm
Transmission: Thirty-six-speed (F & R)
Wheelbase: 112in (2,800mm)
Weight: 12,760lb (5,800kg)

Despite its new range of Perkins-powered machines, Landini was still a small fish in a big pond, and was taken over by Massey-Ferguson in 1960. This gave Landin financial security and M-F access to a wide range of crawlers, at which Landini had considerable expertise.

There was a major new range in 1973, the 6/7/8500 line-up with 12+4 transmission, while in 1977 Landini announced Europe's first 100hp four-wheel-

drive tractor with a conventional layout. New orchard tractors appeared in 1982, and a vineyard range in '86. Again, this became something of a Landini speciality, partly due to the large home market, and partly to its existing concentration on smaller machines – these were sold in some markets with Massey-Ferguson badges, while some higher powered Masseys were badged as Landinis.

The mid-range tractors were updated in 1988 by the 60/70/80 series, with 24+12 transmission, and again in 1992. The Blizzard 85 was part of that range, using a Perkins 4.248 non-turbo diesel, giving 80hp and 208lb ft (154Nm). But Landini's top-range tractor for the 1990s was the Legend range, seen here in 123hp 130 form. There were three (110, 123 and 138hp) all powered by (385ci (6.0-litre) Perkins six-cylinder diesels. Four-wheel-drive was standard, with air conditioning on the top two models.

Now with strong links with AGCO, while retaining its own specialist knowledge, Landini looks to be in a good position.

Left: With AGCO backing and Perkins power, plus its own niche in the Italian market, Landini should have a secure future.

Leader Model D

USA

Engine: Water-cooled, four-cylinder
Capacity: 133ci (2,075cc)
Weight: 2,500lb (1,125kg)

Leader's Model D was a small two-plough tractor, introduced in the early 1950s. Its maker, the Leader Tractor Manufacturing Company, was established in 1918, as a successor to the defunct Ohio Tractor Manufacturing Co. One of Leader's early models was a 12-25 similar in layout to the Huber Light Four. That one weighed 5,600lb.

As for the Model D, this weighed less than half as much, an indication of

how far tractor design had come in the intervening years. For motive power, it relied on a Hercules IXB-5 motor, a four-cylinder power unit of 133ci (2,075cc). Like many of the smaller manufacturers, Leader could not afford to make its own engines.

With the Hercules working hard, the Model D could pull two 12-inch ploughs. A belt pulley and PTO were standard.

This Leader should not be confused with the Leader Gas Engine Co, of Grand Rapids.

Below: Leader D was neat, but was squeezed out by more mainstream rivals.

Marshall

UK

Single-cylinder diesel tractors were once something of a European speciality. In Germany, the Lanz Bulldog was a well respected model in the 1920s; France had Vierzon and Italy Landini; in Hungary, the HCSC was a Lanz built under licence. Britain had the Marshall.

Marshall built steam traction engines, and had been doing so since 1848 in the town of Gainsborough in Lincolnshire, at the heart of a large agricultural region. Like its rivals, Marshall could see that the writing was on the wall for steam traction, and in 1908 unveiled a petrol-paraffin tractor, though steam power was the dominant product right into the late 1920s. It was then that a Lanz was brought into the Marshall factory. The result was the Marshall

15/30, with a big single-cylinder two-stroke diesel engine.

Over the years, it acquired quite a following, uprated as the 18/30 and later joined by the smaller 12/20. In 1945, it reappeared as the Series One, updated into Series II and III through the early 1950s. But the multi-cylinder opposition was forging ahead, so Marshall bought in a Leyland six-cylinder engine for the MP6 tractor. Three years later, it stopped making agricultural tractors, but returned in the 1980s, having bought the rights to Leyland tractors. The end of Marshall tractors came in 1990.

Below: Marshall attempted to keep up with multi-cylinder rivals, but failed.

Marshall Series II

UK

Specifications for Series II (1947)
Engine: Water-cooled, single-cylinder diesel
Bore x stroke: 6.5 x 9.0in (163 x 225mm)
Capacity: 298ci (4,656cc)
PTO power: 40hp @ 750rpm
Transmission: Three-speed
Speeds: 2.75-6.0mph, 9mph opt (4.4-9.6km/h, 14.4km/h)
Weight: 6,500lb (2,925kg)

For many, the classic Marshall tractor was the pre-war 12/20, renamed the M in 1938. In the late 1940s and early '50s Marshall attempted to keep up with multi-cylinder rivals by uprating its single-cylinder into the 50hp class. But this was asking too much of an ageing design, which by then was simply outdated. But that was all in the future in 1945, when Marshall announced the Series

One. This was really the pre-war M, but restyled for a more modern appearance, and more power thanks to the rated engine speed increased from 700rpm to 750. Mechanically, it shared much with the original 15/30 of 1930: a two-stroke low-revving horizontal cylinder with a long stroke, driving through a three-speed gearbox. Unlike its 'semi-diesel' rivals, the Marshall was not started by the blowlamp and hot bulb method, but with ignition papers and – later – cartridge starting. An electric start was offered for the last few years. In fact, with no hot bulb, no glowplug or gasoline starting system, simplicity was one of the Marshall's key selling points.

The Series II pictured here appeared in 1947, with (according to *Farming & Implement News*), 'a new and improved braking system, better bearings, better engine cooling and larger rear tyres as standard.'

Below: Distinctive exhaust stack and narrow body – typical Marshall.

Marshall Series III

UK

Engine: Water-cooled, single-cylinder diesel
Bore x stroke: 6.5 x 9.0in (163 x 225mm)
Capacity: 298ci (4,656cc)
PTO power: 40hp @ 750rpm
Transmission: Three-speed
Speeds: 2.6-4.9mph, 6-11.3mph opt (4.2-7.8km/h, 9.6-18.1km/h)
Weight: 6,650lb (2,993kg)

If earlier Marshalls shared a weakness, it was the transmission. The M suffered from cracking of the transmission housing, especially when winches were fitted, and despite increased bearing sizes the Series II was prone to gearbox failures as well.

This was addressed in the Series III with a beefed up final drive – there was a double crown wheel, with reduction gearing now taken care of in housings on either side of the main gearbox. These housings also acted to strengthen the casing, in a bid to defeat the cracking problem. There were engine improvements too – more fuel and coolant capacity, and changes to lubrication, piston and fuel pump. However, the Series III still had difficulty in maintaining its rated 40hp output. It was more important than ever that this was achieved, with mass-produced multi-cylinder diesel tractors from Fordson and Nuffield undercutting Marshall on price.

They finally made it with the Series IIIa of 1952. A pressurised cooling system, more efficient fuel pump and injectors and wide piston top ring now ensured that 40hp could be sustained. In fact, in a serious effort to modernise the Marshall, an electric start was offered, as well as an Adriolic hydraulic lift and three-point linkage. But none of this was enough to cope with the cheaper opposition. A Series IV was planned, and even

Marshall MP6

UK

Engine: Water-cooled, six-cylinder diesel
Power: 70hp
Drawbar pull: 10,000lb (4,500kg)
Transmission: Six-speed

By the mid-1950s, it was clear that big single-cylinder diesels were being rapidly overtaken by the multi-cylinder opposition. This was serious for Marshall, being completely dependent on this type of machine. Landini faced the same problem at the same time, and adopted Perkins diesels. Marshall turned to Leyland.

The new MP6, announced in 1954, could hardly have been more different from the big singles that the Gainsborough factory had produced for 25 years. It used a six-cylinder Leyland truck engine, the UE350, and added a six-speed gearbox as well. On paper it looked promising – the MP6 had an impressive drawbar pull of 10,000lb (4,500kg) – but in practice the MP6 was too big, powerful and expensive for the British home market. Not many MP6s were made, and of the 197 that did leave the works, a mere ten were sold at home. The rest were exported all over the world, notably to the West Indies for sugar cultivation.

This wasn't enough to keep Marshall in the tractor business, and it soon decided to concentrate on crawlers.

Above: Despite improvements, the Series III looked outdated.

supercharging was considered, but 1957 saw the end for Marshall's big single.

Above: Six-cylinder MP6 failed to save Marshall's tractor business.

Massey-Ferguson

UK/USA

Massey-Ferguson grew out of a merger between Massey-Harris and Harry Ferguson in 1953. It wasn't a promising start – M-H built an increasingly outdated line of tractors, while Ferguson was attempting to go it alone after falling out with Ford. But between them, they had the capacity to build a lot of tractors, with factories in the USA, Canada, England and France plus (and this is the clincher) Harry Ferguson's three-point hitch.

The new Massey-Ferguson took over diesel engine manufacturer Perkins in 1959, just as the tractor market was going diesel in a big way. Landini of Italy was bought the year after. And in 1964/5, a new line of rationalised,

worldwide-specification tractors, the 'Red Giants' were launched. By the mid-1980s, M-F could claim to be the largest tractor maker in the Western world. When it was taken over by AGCO in 1994, it more than doubled its new owner's turnover. As part of the AGCO group sitting alongside Allis-Chalmers, Deutz and White, it was inevitable that some rationalisation would take place. But today Massey-Ferguson remains one of the world's leading brands of tractor.

***Below:* Two famous names, one famous brand – Massey-Ferguson.**

Massey-Ferguson 50 UK/USA

Specifications for 50 diesel (1961)
Engine: Water-cooled, three-cylinder
Bore x stroke: 3.6in x 5.0in (90 x 125mm)
Capacity: 153ci (2,387cc)
PTO power: 38.3hp@ 2,000rpm
Drawbar power: 32.4hp
Transmission: Six-speed
Speeds: 1.3-14.6mph (2.1-23.4km/h)
Fuel consumption: 13.42hp/hr per gallon
Weight: 3,933lb (1,770kg)

When Ferguson merged with Massey-Harris in 1953, it soon became clear that there would have to be some serious rationalisation. Both firms had a complete range of small tractors, of which the Fergusons were superior. Ferguson also had new advanced larger machines on the drawing board. But at first both lines were kept going, with some machines repainted and rebadged to suit – some M-H Ponys, for example, were painted grey and sold by Ferguson dealers.

This confused strategy didn't work – it made life more difficult for the dealers, while customers weren't sure whether they were buying a genuine Ferguson or a Massey-Harris. It also failed to take advantage of M-F's huge economies of scale. This all changed in 1957. The whole line-up was badged as Massey-Fergusons, in the new corporate colours of red bodywork with a grey chassis.

So the M-F TO-30 was really a Ferguson TO-30 in the new colour scheme. The

Massey-Ferguson 95 Super UK/USA

Engine: Water-cooled, four-cylinder
Bore x stroke: 4.25 x 5.0in (106 x 125mm)
Capacity: 426ci (6,646cc)
PTO power: 75hp @ 1,300rpm
Drawbar power: 67hp
Transmission: Five-speed
Speeds: 3.1-17.1mph (5.0-27.4km/h)
Weight: c8,000lb (3,600kg)

The newly unified Massey-Ferguson was keen to market a complete range of tractors as quickly as possible. In the small and mid-range, they were well covered, but had nothing bigger than the 50hp M-F 65. The short-term answer was to buy someone else's tractor and bolt a Massey-Ferguson badge onto it. So that's what they did.

Minneapolis-Moline was making something suitable, the big Gvi, with its massive 425ci (6,630cc) six-cylinder diesel. Mechanically, the Massey-Ferguson 95 Super was the same machine, but in M-F colours. With a five-speed transmission, 75hp at the belt and a weight of around 8,000lb 93,600kg), it gave the company the right sort of machine for the American wheatlands. It evidently worked, for M-F went on to buy 500 Model 990 tractors from Oliver – the 60hp Massey-Ferguson 98 was born.

But it would never be the ideal solution – buying someone else's tractors is invariably more expensive than building your own, and these rebadged Olivers and Minneapolis-Molines were never intended as any more than stop-gaps. Sure enough, in 1959 M-F announced its own 60hp tractor, the 88, a 4-5-plough

Above: Massey-Ferguson 50 had first been sold as a Massey-Harris.

50 pictured here was a stretched version of Ferguson's TO-35, which initially had been sold as an all-red Massey-Harris 50 or all-grey Ferguson 40. A 2-3 plough tractor, it started off with a Continental 134ci gasoline engine, later adding LPG and diesel options. The diesel used a Perkins three-cylinder motor of 38hp.

Above: M-F 95 was really a rebadged Minneapolis-Moline.

row-crop tractor with eight-speed transmission and four-cylinder gasoline or diesel power. That was followed by the 68hp M-F90. To fill in until the new line of Red Giants arrived, the M-F97 was another rebadged Minneapolis-Moline, this time the 100hp G705.

Massey-Ferguson 135

UK

Specifications for 135 (1965)
Engine: Water-cooled, four-cylinder
Capacity: 152ci (2,371cc)
PTO power: 35.4hp @ 2,000rpm
Drawbar power: 30.5hp
Transmission: Twelve-speed
Speeds: 1.4-19.8mph (2.2-31.7km/h)
Fuel consumption: 9.02hp/hr per gallon
Weight: 3,565lb (1,604kg)

A familar sight on English farms, the little 135 was the smallest of the new line of 100-series Red Giants announced at the Smithfield Agricultural Show in 1964. Together, they were an important step forward for Massey-Ferguson, representing as they did a complete renewal of the tractor range. Thousands of man-hours had gone into designing and testing this new range, which was probably the most important in Massey-Ferguson's history. More than that, it was a rationalised range, of which the smaller

tractors were built in England and France, the larger ones in North America. And finally, to simplify and further rationalise matters, they were to a worldwide specification, with only minor differences between different national markets.

Not that the 135 was all-new. It had the same power unit options – Continental 134ci (2,090cc) gasoline or Perkins 152ci (2,371cc) three-cylinder diesel – as the 35 it replaced. Although the 135 was made at M-F's British plant in Coventry, some were exported across the Atlantic – hence the gasoline option, for in Europe by the early 1960s diesel power had all but taken over. It also retained the 35's beam front axle, though the larger 150 and 165 had a different front end to accommodate row-crop equipment. The transmission was new, a six-speeder with two reverse, though these ratios were doubled it you paid extra for 'Multipower.' Finally, there was new squared-off styling common to the entire 100-series range, underlining the fact that this was a new generation.

Below: Most 135s were Perkins-powered, but a few gasoline versions emerged.

Massey-Ferguson 150

UK

Specifications for 150 diesel (1965)
Engine: Water-cooled, three-cylinder
Bore x stroke: 3.6in x 5.0in (90 x 125mm)
Capacity: 152ci (2,371cc)
PTO power: 37.8hp @ 2,000rpm
Drawbar power: 33hp
Transmission: Twelve-speed
Speeds: 1.4-18.6mph (2.2-29.8km/h)
Fuel consumption: 15.4hp/hr per gallon
Weight: 4,805lb (2,162kg)

Just as the 135 was a straight replacement for the M-F 35, so the new 150 did the same for the old Model 50. Engine and transmission options were the same as for the smaller tractor, and the hydraulics were unchanged from the 50, except that the control was repositioned on the right side of the gearbox. There was also a new feature named Pressure Control, which allowed for weight transfer from implements, plus auxiliary hydraulics for

operating equipment off the tractor. This was soon made standard on the 165 and 175, though remained an option on the smaller machines. By the late '60s, gasoline engines were a minority part of the European tractor market,but they were still mainstream enough in America to keep this option open. In 1969, M-F dropped the Continental engines, adopting gasoline versions of the Perkins diesels.

In fact, the 150, so similar in specification to the 135, was later dropped. But there were plenty of other tractors to choose from. If the 135 wasn't small enough, there was the 130. This was made in France, and the smallest M-F sold in the UK, using a Perkins 107ci (1,669cc) four-cylinder diesel as standard. The French arm of M-F also produced vineyard and orchard versions of both this and the 135 – the old 35, too, had been made in France as well as England. The French connection actually went back to the Massey-Harris days, when versions of the M-H Pony were built in France with Simca or Peugeot gasoline engines, or a Hanomag diesel.

Below: Well-worn M-F 152, very similar to the 135, and later dropped.

Massey-Ferguson 165

UK

Specifications for 165 (1965)
Engine: Water-cooled, four-cylinder
Bore x stroke: 3.58in x 4.38in (90 x 110mm)
Capacity: 176ci (2,746cc)
PTO power: 46.9hp @ 2,000rpm
Drawbar power: 39.9hp
Transmission: Twelve-speed
Speeds: 1.3-18.5mph (2.1-30km/h)
Fuel consumption: 9.43hp/hr per gallon
Weight: 5,005lb (2,252kg)

Massey-Ferguson's 165 represented the top end of its small tractor range, sharing many of the same features with the 135 and 150. It did have a bigger engine though, the same 50hp 203ci (3,167cc) Perkins diesel used in the 65 it replaced, plus a 176ci (2,746cc) Continental for those who still preferred gasoline. It was joined by a more powerful 175, with another Perkins four-cylinder, but of 236ci (3,682cc) capacity – it was an engine

designed specifically for this tractor. In 1971, the 175 was itself replaced by the more powerful still 178, this time with a 248ci (3,869cc) Perkins.

Late that year, the 100-series diverged into two separate lines. For those who wanted a basic, no-frills value-for-money tractor, the 'Standard Rig' 135, 165 and new 185 had no Multipower or power steering. If your purse was deeper, there was the 'Super Spec' with a longer wheelbase than the standard machines. This was to allow the fitting of a more spacious safety cab – the first cabs had restricted access – and also brought better weight distribution, allowing the use of heavier implements without the need for front weights. Super Spec tractors also had a Multipower transmission, independent PTO, a spring suspension seat and high capacity hydraulic pump. The top model added power steering and power-adjusted wheel track. In fact, the Super Spec tractors were so different from their basic brethren that they had different names as well: 148 (135 equivalent), 168 (165) and 188 (185).

Below: This was M-F's top small tractor in 1965.

Massey-Ferguson 1135

UK/USA

Specifications for 1135 diesel (1973)
Engine: Water-cooled, six-cylinder, turbo
Bore x stroke: 3.88in x 5.0in (97 x 125mm)
Capacity: 354ci (5,522cc)
PTO power: 120.8hp @ 2,200rpm
Drawbar power: 102.4hp
Transmission: Twelve-speed
Speeds: 2.2-16.9mph (3.5-27.0km/h)
Fuel consumption: 13.22hp/hr per gallon
Weight: 13,550lb (6,098kg)

Massey-Harris, and now Massey-Ferguson, had traditonally been stronger on small and mid-sized tractors than big ones – that's why they bought-in machines from Oliver and Minneapolis-Moline to fill a gap in the range.

The market for such tractors was still mainly in the USA and Canada in the 1960s, and the expertise for building such machines lay there as well. So that's where the new Massey-Ferguson 1100 and 1130 were made. They didn't have four-wheel-drive or articulation (though M-F's contender for that market was already on the drawing board), being instead high-power conventional machines. The 1100

came with a 354ci (5,522cc) six-cylinder Perkins diesel of 90hp, while the 1130, introduced in 1968, used a turbocharged version ofthe same engine, with 120hp – it was in fact the first turbocharged Massey-Ferguson. The 1135 pictured here is a slightly updated version of that tractor, with the same engine. Both these new big tractors boasted an Advanced Ferguson system with more sophsticated hydraulics, finer control and the ability to take larger implements. And in keeping with the times, driver comfort was given new priority.

But big and powerful though these might be by Massey-Ferguson standards, they were nothing special compared to the American opposition. A 135hp 1150, powered by a turbo-diesel V8, arrived three years after the 1130. But what M-F really needed was a big 4x4 tractor to suit the massive fields of North America. The Canadian-built 1200 was its first, powered by the now-familiar Perkins 354ci (5,522cc) six, while the 1500 and 1800 which followed used Caterpillar V8 diesels, with 153hp and 179hp respectively. The 4x4 range was replaced by the 4000 series in 1975, but M-F's biggest tractors were all rationalised out of existence after the AGCO takeover.

Below: The six-cylinder turbocharged 1135 was made in the USA, though still powered by Perkins.

Massey-Ferguson 3095 UK/FRANCE

Specifications for USA model, 1984
Engine: Water-cooled, six-cylinder, turbo
Bore x stroke: 3.875 x 5.0in (97 x 125mm)
Capacity: 354ci (5,522cc)
PTO power: 91.5hp
Drawbar power: 73.4hp
Transmission: Sixteen-speed
Speeds: 1.3-18.2mph (2.1-29.1km/h)
Fuel consumption: 14.3hp/hr per gallon
Weight: 12,935lb (5,821kg)

By the late 1970s, just as M-F's American arm had specialised in the bigger tractors, so the Europeans had their own division of labour. The Beauvais plant in France tended to produce high-tech mid-size machines, while smaller tractors carried on from the Banner Lane factory in Coventry, which could trace its tractor production back to the early Ferguson days.

The 500 series for example, had been unveiled in 1976 as a more modern

supplement to the ageing but popular 200 series, with low-noise cab and ergnomic controls. This was a Banner Lane tractor, but the 600 which replaced it was an English/French joint project. Meanwhile, the new top-range 2000-series was being designed and built in Beauvais. Announced in 1979, the 2640, 2680 and 2720 covered the 110-150hp range, with all three machines using a variant of the faithful Perkins six, the most powerful of which now used an intercooler to produce 147hp. All had a high-tech push-button transmission which allowed full-power range changes, and offered 16 forward speeds, 12 reverse.

The 3095 pictured here represents the 3000-series, which arrived in 1986 to slot in underneath the 2000 (so why didn't they call it the 1000 – who knows?) A range from the 71hp 3050 to the 107hp 3090 (the 3095 shown here soon joined) was offered, all with electronic control of the three-point linkage. On higher spec versions, you also had Autotronic, with automatic operation of the differential lock, front-wheel-assist and PTO.

Below: Top of the 1986-on 3000 series, the 3095.

Massey-Ferguson 3630 UK

Specifications for USA model, 1984
Engine: Water-cooled, six-cylinder, turbo
Bore x stroke: 3.875 x 5.0in (97 x 125mm)
Capacity: 354ci (5,522cc)
PTO power: 108hp @ 2,400rpm
Drawbar power: 90hp
Transmission: Sixteen-speed
Speeds: 1.3-18.2mph (2.1-29.1km/h)
Fuel consumption: 16.7hp/hr per gallon
Weight: 13,100lb (5,895kg)

In 1987, Massey-Ferguson replaced the 2000-series altogether, with higher-powered versions of the 3000. The change brought slightly more power as well, with 113hp, 130hp (turbo) and 150hp (intercooler). Three years on all the 3000s were themselves replaced by the 3600 range. The 3000 was replaced with a wider range of 3600s, which in their higher-powered guises also replaced the 2000.

The lower powered 3600s, including the 3630 pictured here, used Perkins latest 1000-series six-cylinder direct injection diesel. Direction injection was more efficient and indirect, and in this particular engine was combined with something called the Quadram system. This was a combustion bowl in the piston crown, with four lobes cast into it to optimise mixing of the fuel/air charge. According to Perkins, this made combustion faster and evened out combustion pressure peaks, which it said improved

Massey-Ferguson 3690 UK

Specifications for 8245 (2002)
Engine: Water-cooled, six-cylinder, turbo-diesel
Bore x stroke: 4.3 x 5.3in (108 x 134mm)
Capacity: 452ci (7.4 litres)
PTO power: 160hp @ 2,000rpm
Torque: 600lb ft (443Nm)
Transmission: Sixteen-speed
Speeds: 1.5-24.9mph (2.4-40.1km/h)
Weight: 18,500lb (8,392kg)

It was 1992 when the 3600-series received its final major change, before being replaced altogether by the 6100/8100. It was Dynashift, the latest development of ever more sophisticated transmissions which were needed partly to cope with the higher power of modern tractors, partly because of the ever more specialised roles they were taking on and partly because electronic control made it all possible.

Dynashift used a four-speed epicyclic gear system mounted ahead of the standard 8-speed gearbox. This gave 32 speeds in forward or reverse, 24 of which could be selected without the clutch. But the sophstication was such that if the driver made a bad shift by pressing the wrong button, the electronics would block it, thus protecting the engine and transmission from mechanical mayhem if, say a low-range reverse gear was selected at top forward speed!

Another use of electronics in the 3600 was Datatronic, an addition to the Autotronic which allowed automatic operation of front-wheel-assist, the

Above: Lower powered 3600s used Perkins six-cylinder diesels.

torque at low speeds. The highest powered 3600 actually used a Valmet engine – it was rare for a Massey-Ferguson to fit a non-Perkins diesel (Perkins being part of the corporate family) but it sometimes had no choice in the higher powered applications, for which Perkins had no suitable engine.

Above: Final 3690 had Dynashift, a sophisticated 32-speed transmission.

PTO and differential lock. Datatronic added a a digital display of all vital functions in the cab. It also had wheel-slip control, whereby the driver could limit the degree of slip to maximise traction. Specifications are given for the 8200 series, the 2002 equivalent of the 3690.

Massey-Harris

CANADA

Massey-Harris was Canada's most successful tractor manufacturer, and as part of Massey-Ferguson (itself now owned by AGCO) the name lives on today. Daniel Massey and Alanson Harris, both makers of mowers and binders, merged in 1891. They were keen to expand their product line, though it was ten years before gasoline engines were introduced.

At first, Massey-Harris didn't design its own tractor, but distributed the Minneapolis-built Big Bull from 1915. The Big Bull was not a success, but Massey-Harris had already commissioned one Dent Parrett to design a machine for them. This was the first Massey-Harris badged tractor and was built up to 1923. Not until 1930 did the Canadians produce their own tractor, the intriguing four-wheel-drive General Purpose, but a high price and various flaws kept sales

low. But M-H persevered with a range of conventional, slightly outdated machines inherited from the Wallis/JI Case Plow Works (not to be confused with the JI Case of Case-International), and its own smaller tractors such as the 101 and Pony.

These still lacked advanced features like live hydraulics and a three-point hitch, but a solution was on the horizon. Harry Ferguson, having fallen out with Ford, was looking for a manufacturing partner, and in 1953 Massey-Harris-Ferguson was born. Four years later, the company was restructured into Massey-Ferguson, which marked the end of the Massey-Harris name, and its red and yellow tractors.

Below: Massey-Harris was independent for over 60 years.

Massey-Harris No2

CANADA

Specifications for No2 (1921)
Engine: Water-cooled, four-cylinder
Bore x stroke: 4.25 x 5.5in (106 x 138mm)
Capacity: 287ci (4,477cc)
PTO power: 22hp @ 1,000rpm
Drawbar power: 12hp
Transmission: Two-speed
Speeds: 1.75 and 2.4mph (2.8 and 3.8km/h)
Weight: 5,200lb (2,340kg)

Massey-Harris' second bid for a slice of the tractor market came in three sizes: the 12-25hp No1, 12-22hp No2 (pictured here) and 15-28hp No3. All were designed in Illinois by Dent Parrett on behalf of M-H, but at least some of them were made in Canada.

They were simple machines, basically just a steel frame with a Buda four-cylinder engine mounted crossways – Nos 1 and 2 had the radiator mounted in-line, but No3 mounted it conventionally, facing the front. All of them drove through a two-speed + reverse gearbox, and the chief distinction between the first two was that No2 had enclosed rear wheel gears. Despite using the same 287ci (4,477cc) Buda engine, the No2 was slightly derated to 22hp at the belt (No1, 25hp) though with the same 12 drawbar hp. Either would run on gasoline or kerosene, as would the larger No3, which used a 397ci (6,193cc) Buda.

Unfortunately, Parrett had designed the tractor before 1914, and the

Massey-Harris 20/30

CANADA

Specifications for 20-30 (1931)
Engine: Water-cooled, four-cylinder
Bore x stroke: 4.375 x 5.75in (109 x 144mm)
Capacity: 346ci (5,398cc)
PTO power: 30hp @ 1,050rpm
Drawbar power: 20hp
Transmission: Two-speed
Speeds: 2.75 and 3.3mph (4.4 and 5.3km/h)
Weight: 4,381lb (1,971kg)

Three years after giving up on the Parrett tractors, Massey-Harris tried again. It still didn't feel ready to design its own tractor in-house, so negotiated to distribute Wallis tractors in Canada. The scheme evidently worked, as another two years on M-H took over the company that made Wallis machines. It wasn't long before they were wearing Massey-Harris badges.

Initially, the range was restricted to the Wallis 15-27, but that was soon uprated into the 20-30, the tractor shown here, and joined by a smaller 12-20. All of these were based on the original Wallis Cub of 1913. This was quite advanced for its time, with a light, rigid unit construction frame, and M-H carried on building its descendants in the same factory in Wisconsin.

The 20-30 was powered by a 346ci (5,398cc) four-cylinder engine and became Massey-Harris' standard row-crop tractor of the 1930s. It actually came in various guises, all based around the same engine, chassis and transmission. An orchard model offered a low seat, enclosed front wheels

Above: The early M-H was outdated and expensive.

fast pace of tractor technology soon made these Massey-Harris machines look crude and expensive. Why buy one of these when a Fordson was a fraction of the price? M-H abandoned the Parrett tractors in 1923.

Above: 20/30 was an updated Wallis tractor – in-house designs came later.

and full mudguards for the rear wheels, while the 20-30 Industrial had a choice of solid rubber or low pressure pneumatic tyres, plus higher gearing than the field machines. In 1932 it was uprate with more power, and renamed the Model 25, soldiering on until 1939. Although really intended mainly for belt work, the 25 was often used for ploughing as well.

Massey-Harris General Purpose

CANADA

Specifications for General Purpose (1930)
Engine: Water-cooled, four-cylinder
Bore x stroke: 4.0 x 4.5in (100 x 125mm)
Capacity: 226ci (3,527cc)
PTO power: 24.8hp @ 1,200rpm
Drawbar power: 19hp
Transmission: Three-speed
Speeds: 2.2-4.0mph (3.5-6.4km/h)
Weight: 3,861lb (1,737kg)

Until 1930, Massey-Harris had bought-in tractors to sell under is own name. The General Purpose changed all that, being designed and built by M-H in Canada. It was also unique. In some ways, the GP was well ahead of its time, with four-wheel-drive and an articulated chassis, principles which wouldn't enter mainstream tractor design for another 30 years. And yet, as a true general purpose machine, the GP failed to make the grade.

At the time, every tracor maker was keen to produce a tractor that could operate cultivators in row-crop, pull heavy drawbar loads and do belt work – general purpose. The success of the International Farmall showed that there was a huge market for this type of machine. The Massey-Harris offering wasn't one of them: it had a wide turning circle and (a serious drawback for a general

Massey-Harris Challenger

CANADA

Specifications for Challenger (1936)
Engine: Water-cooled, four-cylinder
Bore x stroke: 3.88 x 5.25in (97 x 131mm)
Capacity: 248ci (3,869cc)
PTO power: 27.2hp @ 1,200rpm
Drawbar power: 16.3hp
Transmission: Four-speed
Speeds: 2.4-8.5mph (3.8-13.6km/h)
Fuel consumption: 6.65hp/hr per gallon
Weight: 4,200lb (1,890kg)

After the innovative GP had failed to set the world of row-crop tractors alight, Massey-Harris tried again with a far more conventional machine. The Challenger of 1936 was simply an adapted version of the old 12-20, with the same 248ci (3,869cc) engine(now in overhead valve form) and unit frame.

Instead of the 12-20's standard-tread, it had narrow front wheels and large rear wheels, whose track could be altered by moving them in or out on splined axles. This wasn't a new idea, but it was simple and it worked – in any case, Massey-Harris desperately needed a competitive row-crop tractor with features like this, simply because all its rivals already offered them. With the engine uprated to 16-26hp and a reasonable all-up weight of 3,700lb (1,665kg), the Challenger looked promising. M-H's advertising copy writers certainly thought so: 'CHALLENGER. Here's a trim looking tractor as you've ever seen. note the clean, flowing lines – nothing there to obstruct your vision of the field ahead...You'll like this tractor's husky 2-3 plow power – the eagerness with

***Above:* Four-wheel-drive and articulation – the GP was ahead of its time.**

purpose tractor) the track width wasn't adjustable, though it could be ordered in six alternative widths from the factory.

However, it did have some unique selling points. Four-wheel-drive from equal-sized wheels gave it unrivalled traction among wheeled machines; the articulated chassis allowed the rear axle to follow rough ground independently; and weight distribution was carefully thought out, for optimum placing on all four driven wheels. These attributes did secure the M-H a niche in forestry and speciality crops, but at $1,000 it was too expensive for the mass market.

***Above:* Conventional Challenger succeeded where advanced GP failed.**

which it responds to the toughest jobs. Learn the advantages of more power for every job at low cost – make it a point to see the Challenger next time you're in town.'

Massey-Harris 101 Junior CANADA

Specifications for 101 R Junior (1939)
Engine: Water-cooled, four-cylinder
Bore x stroke: 3.0 x 4.39in (75 x 110mm)
Capacity: 124ci (1,934cc)
PTO power: 25.4hp @ 1,800rpm
Drawbar power: 16.4hp
Transmission: Four-speed
Speeds: 2.6-17.4mph (4.2-27.8km/h)
Fuel consumption: 9.75hp/hr per gallon

As World War II began in Europe, Massey-Harris was busy updating its entire line-up of tractors. The 101 came first, in 1938, in power rating fitting between the Challenger and now ageing Model 25. Power unit wise, it was more sophisticated than both, with a smooth Chrysler straight six of 201ci (3,137cc) – the success of tractors like the Oliver 70 had proved that some farmers liked the extra smoothness of six cylinders. The unit frame, which could trace its roots back to the 1913 Wallis

Cub, was discarded, though it is debatable whether the heavy cast-ron frame that replaced it was actually an improvement. There was a four-speed gearbox though, and modern streamlined styling.

The 101 Junior picutured here was really a smaller version of the 101, and was announced the year following. In fact, it was much smaller, a whole 2,000lb (900kg) lighter and powered by a little 124ci (1,934cc) four-cylinder engine in gasoline form only. The motor was bought in from Continental – M-H rarely built its own engines, and would carry on buying from Continental into the 1950s. With 16-24hp, it was a two-plough tractor, filling a gap at the bottom of the Massey-Harris range.

It was later joined by the 101 Junior Twin Power, which had a slightly larger (140ci/2,184cc) version of the Continental engine, plus the Twin Power feature first seen in the Challenger. This was basically a two-speed governor, allowing a higher governed speed for belt work.

Below: Four-cylinder 101 Junior was much smaller than the original 101.

Massey-Harris 101 Super CANADA

Specifications for Super 101 (1941)
Engine: Water-cooled, six-cylinder
Bore x stroke: 3.25 x 4.375in (81 x 109mm)
Capacity: 218ci (3,401cc)
PTO power: 36hp @ 1,800rpm
Drawbar power: 24hp
Transmission: Four-speed
Speeds: 2.4-16.1mph (3.8-25.8km/h)
Fuel consumption: 7.46hp/hr per gall
Weight (steel wheels): 3,805lb (1,712kg)

The Massey-Harris 101 received a boost in 1941. Now with a 218ci (3,401cc) Chrysler engine (still six cylinders of course) it became the 101 Super, though the 201ci (3,137cc) engine was still available as well. This no doubt boosted power significantly (though unfortunately the precise figures aren't available) and there was also a 101 Senior with a 244ci (3,806cc) Continental six-cylinder

unit. That one gave a top speed of 17mph in fourth gear, on rubber tyres. Like the Super, it was rated as a 3-4 plough tractor, but with extra power for belt work.

Both these 101s came with Twin Power, which on the Super 101 allowed 1,500rpm in the first three gears, and 1,800 in top and for the belt. This was a new era of higher tuned tractors, and the Super actually required 70 octance gasoline. But really these higher powered, higher revving engines were gasoline's last hurrah for tractor use. Engine speeds didn't get much higher than this, and diesel was soon to take over as the dominant form of fuel.

In fact, Massey-Harris was one of the first companies to offer diesel in its large and mid-sized tractors, though not just yet. In 1940 it had replaced the Model 25 with the 201, a larger four-plough tractor based on the 101 layout. This came in standard-tread only, with a 242ci (3,775cc) 57hp Chrysler gasoline engine; the similar 202 offered a little more power (60hp Continental) but the top of the range 203 used a 330ci (5,148cc) Continental with 64hp.

Below: Chrysler power for the 101 Super – M-H often bought-in engines.

Massey-Harris 81R

CANADA

Specifications for 81 Rowcrop (1941)
Engine: Water-cooled, four-cylinder
Bore x stroke: 3.0 x 4.375in (75 x 109mm)
Capacity: 124ci (1,934cc)
PTO power: 26hp @ 1,800rpm
Drawbar power: 16.4hp
Transmission: Four-speed
Speeds: 2.4-15.8mph (3.8-25.3km/h)
Fuel consumption: 10.91hp/hr per gall
Weight: 2,895lb (1,303kg)

Just when it seemed as though Massey-Harris was developing a nice, neat, logical numbering system for its tractors – 101, 102, 201, 202, 203 – along came the 81. It appeared in 1941, as a sort of lighter, cheaper alternative to the

101 Junior. It used the same 124ci (1,934cc) Continental gasoline engine as the Junior. The similar 82 as well, but in distillate form, in which case a slightly larger bore of 3.188in (78mm) gave 140ci (2,184cc) to make up for the lower efficiency of distillate, and a lower 5:1 compression ratio – the 81 used a 6.75:1 compression.

Both were two-plough machines, available in standard tread or row-crop (twin front wheel tricycle layout). And both continued the Twin Power feature, though it wasn't called that any more, which allowed a higher governed speed of 1,800rpm for belt work or top gear driving. The 81/82 were updated as the Model 20 soon after the war, still at 2,700lb (1,215kg) a lightweight machine. Up half a class was the Model 30, which replaced the Super 101 and had a new five-speed transmission – rated power of that one was 20-30hp. It still used a Continental engine, the 162ci (2,527cc) four.

Below: Two-plough 81 was a cheaper alternative to the 101 Junior.

Massey-Harris 102 Junior

CANADA

Specifications for 102 Junior (1946)
Engine: Water-cooled, four-cylinder
Bore x stroke: 3.44 x 4.38in (86 x 110mm)
Capacity: 162ci (2,527cc)
Power: n/a
Transmission: Four-speed
Speeds: 2.2-10.1mph (3.5-16.2km/h)
Wheelbase: 78in (1,950kg)

This tractor, the Massey-Harris 102 Junior, was a slightly more powerful version of the 101 Junior, using a 162ci (2,527cc) Continental power unit in place of the 101's 124ci (1,934cc), and was announced in 1946. There was a bewildering array of variants. The 102G and GS Junior were the standard-tread versions, the G only listed with steel wheels, the GS with pneumatic rubber tyres, though the

GS was also described as a 2-3 plough tractor and the G as 2-plough. There were row-crop versions of both of these, the 102 Junior Rowcrop G and GRC. There was also an ochard model.

Take off the 'Junior' tag, and the 102 started to get serious. As the 102G Twin Power, it was endowed with a six-cylinder Continental engine of 226ci (3,526cc). As the name suggested, it also had M-H's well-proven twin-speed governor to allow 1,800rpm at the belt.. For 1944/45 there were G and GS versions of this one, with either the 226 (3,526cc) or a 244ci (3,806cc) Continental, depending on who you ask. Confused?

Fortunately, things were soon simplified when the whole M-H line-up was updated – the 102 was replaced by the 44, one of the company's most successful machines.

Below: 'Twin Power' denoted a two-speed governor on the 102 Junior.

Massey-Harris Pony

CANADA

Specifications for Pony (1948)
Engine: Water-cooled, four-cylinder
Bore x stroke: 2.38 x 3.5in (60 x 88mm)
Capacity: 62ci (967cc)
PTO power: 10.4hp @ 1,800rpm
Drawbar power: 8.3hp
Transmission: Three-speed
Speeds: 2.7-7.0mph (4.3-11.2km/h)
Fuel consumption: 9.02hp/hr per gallon
Weight: 1,890lb (851kg)

Announced in 1947, the little Pony was Massey-Harris' attempt to build a small one-plough tractor for small farmers, market gardeners and tobacco growers. Interestingly, despite the company's Canadian base, it was the only M-H tractor actually built in Canada in any numbers, apart from some of the Parrett types.

Most Massey-Harris badged tractors were made in the USA, England or France.

Weighing less than 1,900lb (855kg), the Pony produced 8 drawbar hp, and 10 at the belt, and for the American market used a four-cylinder Continential engine of just 62ci (967cc). It shared the same familiar styling of the bigger M-Hs and came with an adjustable front wheel width and high-clearance front axle options. What it didn't have until later (in the No14 Pony) was a hydraulic coupling, and it's thought that less than 100 Ponys were so equipped.

The Pony was also assembled in France as the model 811 later with a 78ci (1,221cc) Simca engine replacing the Continental. That was replaced by the Pony 812, which later switched to a Peugeot unit, and then the Pony 820, using both Simca and Peugeot gasoline engines as well as a 65ci (1,021cc) Hanomag two-stroke diesel.

Below: My little Pony – 967cc one-plough machine with 10hp.

Matbro TR200

UK

Specifications for TR250 (1997)
Engine: Water-cooled, four-cylinder, turbo
Capacity: 256ci (4.0 litres)
Power: 106hp
Transmission: Four forward, three reverse
Lifting weight: 2.5 tons

Henry Ford pioneered the idea that one tractor should be able to do any job around the farm. How times have changed. Few farms now rely on one machine, and many buy more specialised tractors suited to different tasks.

The Matbro is an excellent example. You wouldn't use it for ploughing, but that's not what it's designed for – it is a yard shunter, pure and simple. So it has

to be manoeuvrable – neither of the axles steer, but the whole machine is articulated to give a tight turning circle in confined space. It's really a super-forklift for the farm, able to lift up to 2.5 tons at a time from its front loader. There's four-wheel-drive and a torque convertor transmission, while the top of the range TR250 has a Clark powershift transmission with four forward speeds and three reverse, with electronic selection.

Power comes from a Perkins 4.0 litre diesel, with or without turbos, in 75, 96, 106 and latterly 114hp forms. In 2001, Matbro, based in the English Cotswolds, was taken over by the US Terex Corporation. Production of the renamed Terex TM200R and 250R was moved to Manchester late that year.

Below: Matbro TR200, pictured in Glastonbury, southern England.

Mercedes-Benz MB-trac 1300 GERMANY

Specifications for Unimog 30 Diesel (1957)
Engine: Water-cooled, four-cylinder
Bore x stroke: 2.95in x 3.94in 74 x 99mm
Capacity: 108ci (1,685cc)
PTO power: 27.3hp @ 2,550rpm
Drawbar power: 20.6hp
Speeds: 0.72mph-33.0mph (1.2-53km/h)
Fuel consumption: 11.42hp/hr per gallon
Weight: 4,979lb (2,263kg)

Not many people associate truck and luxury car maker Mercedes Benz with tractors, but its agricultural roots go back to 1919, when Benz first unveiled two Land Traktor's of 40 and 80hp. A diesel engined tractor (one of the first) followed a few years later, the 30hp twin-cylinder S7. Initially a tricycle tractor (albeit with outriggers to improve stability) it was soon replaced by the four-

wheel BK. But the company seemed sold on diesel power, producing the single-cylinder OE diesel in the 1930s. Its single cylinder was mounted horizontally and produced 20hp.

But Mercedes is best known for the ubiquitous Unimog, originally produced by another German firm, Boehringer, but later sold to M-B. Best described as half-tractor, half-truck, this unique vehicle has now been in production for over 40 years. Belying its truck-like appearance, the Unimog has four-wheel-drive, lots of ground clearance and excellent towing ability. In the early years, Mercedes publicity described it as a versatile field tractor as well as road hauler. Since then, it's made good that promise. More recently Mercedes has branched into the pure tractor market. The MB-trac 1500 is interesting – with three-point hitches at both ends, it can be used to pull or push implements as required.

Below: MB-trac 1300 was part tractor, part truck.

Minneapolis-Moline/ Twin City

USA

Minneapolis-Moline was actually a combination of three companies. The Minneapolis Threshing Machine Company, the Moline Plow Company and Minneapolis Steel & Machinery all merged in 1929. Moline started out in 1852 making farming implements, and Minneapolis Threshing, as the name suggested, concentrated on threshers, plus steam traction engines and some gasoline tractors. But M-M's real tractor heritage came from Minneapolis Steel & Machinery, established in 1902 and building tractors from 1910, at first under contract for Case and Bull. The firm's own machines were built under the 'Twin City' name.

These three made up Minneapolis-Moline. Although most M-M tractors

were solid, convetntional machines, there were also some startling innovations, such as the striking looking UDLX Comfortractor, the car-like 16-30 or 16-valve 12-20. It wasn't always a complete range though, and the company later tended to neglect the lower end of the market. In 1963, it was purchased by White, which heralded a sad decline through the '60s and early '70s. In the words of author P. W. Ertel (The American Tractor), 'Minneapolis-Moline was now....unceremoniously relegated to the sidelines.' After a few years of badge-engineering, White finally dropped the M-M name in 1974.

Below: A merger of three companies produced Minneapolis-Moline.

Minneapolis Twin City 60-90 UK

Specifications for 60-90 (1915)
Engine: Water-cooled, six-cylinder
Bore x stroke: 7.25in x 9.0in (181 x 225mm)
Capacity: 2,230ci (34,788cc)
PTO power: 90hp
Drawbar power: 60hp
Transmission: Single-speed
Weight: 28,000lb (12,727kg)

Early Twin Cities were huge, and in this they were following conventional thinking. In those pre-Fordson, pre-Farmall days, many early tractors were simply gasoline versions of steam traction engines. Big farms could afford them for ploughing, or powering large threshers, but they were far from the simple, cheap do-it-all tractor which small farmers could afford.

So the 60-90 of 1913 (it actually started out as a 60-110, but was downgraded in its first year) was a 14-ton monster that looked exactly like the steam traction engines it competed with. Even the giant 100-gallon fuel tank looked like a truncated steam boiler! The difference was in the six-cylinder gasoline engine, with similarly massive dimensions – each piston measured over seven inches across and rumbled up and down a stroke of nine inches, which gave a capacity of over 2,200ci (34.3 litres).

With 90hp at the belt, it had no problem powering the largest thresher, but could also be found ploughing or even helping build roads. A single forward gear gave a speed of 2mph and a large canopy served to protect the engine, which

Minneapolis Twin City 16-30 USA

Engine: Water-cooled, four-cylinder
Bore x stroke: 5.0in x 7.5in (125 x 188mm)
Capacity: 588ci (9,172cc)
PTO power: 30hp
Drawbar power: 16hp
Transmission: Two-speed
Weight: 7,800lb (3,545kg)

The 16-30, unveiled in 1917, was quite a departure for Minneapolis – in fact, for the tractor industry in general. Until then, most tractors had left the driver and engine exposed to the elements, but the 16-30 was low and streamlined, enclosing the engine and part of the driver's platform in sheet metal. It was also much smaller than giants like the 60-90, weighing a comparatively lightweight 7,800lb (3,545kg) and using the company's own 588ci (9.2 litres) four-cylinder engine. In appearance as well as design, the 16-30 made a big step forward towards the modern tractor layout, but it was another decade before those low lines and enclosed bodywork became the norm.

There were other new features too. An electric starter and lights could be ordered, and there was a K-W high-tension magneto. Hyatt roller bearings were used throughout, a great improvement over the bronze plain bearings used previously. Unfortunately, the 16-30 also suffered from starting problems, and only 702 were built. But it didn't put Minneapolis off advanced tractors – two years after the 16-30 it announced the little 12-20, an up to the minute four-cylinder tractor with four valves per cylinder.

Above: Now that's big – 34 litres, 14 tons, 90hp.

otherwise was open to the elements. The 60-90 was still listed until 1920, though not many were made. At over $4,000 apiece in 1915, that's hardly surprising – M-M was lucky to have the contracts with Bull and JI Case to keep it afloat.

Above: Electric starting, full bodywork – the advanced but flawed 16-30.

Minneapolis Twin City 17-28 USA

Specifications for Twin City 17-28 (1926)
Engine: Water-cooled, four-cylinder
Bore x stroke: 4.25in x 6.0in (106 x 150mm)
Capacity: 340ci (5,304cc)
PTO power: 30.9hp @ 1,000rpm
Drawbar power: 22.5hp
Transmission: Two-speed
Fuel consumption: 9.58hp/hr per gallon
Weight: 5,895lb (2,680kg)

The 17-28 had its roots in the 12-20 of 1919 – it was really just an uprated version of the same tractor. The 12-20 – despite costing twice as much as a Fordson – was a good seller, and by the end of its first year nearly 3,000 had found homes. So successful was it that Minneapolis was able to claim fourteen years later that its successor was still selling as well.

With four valves per cylinder and unit construction (that is, the engine and transmission were stressed members, forming the chassis themselves) it was certainly advanced. In a 1934 catalogue, the company emphasised this point, but also (maybe with one eye on a conservative marketplace) long life and reliability – Minneapolis claimed that its tractors lasted a good three years longer than the opposition:

'Three extra years is the reputation of all Twin City tractors. The Twin City 17-28 is truly the pioneer of modern tractor design and construction. Of course, it has every worthwhile modern improvement. Owners claim the lowest cost

Minneapolis-Moline Twin City 21-32 USA

Specifications for Twin City 21-32 (1926)
Engine: Water-cooled, four-cylinder
Bore x stroke: 4.5in x 6.0in (113 x 150mm)
Capacity: 382ci (5,959cc)
PTO power: 35.9hp @ 1,000rpm
Drawbar power: 31.1hp
Transmission: Two-speed
Speeds: 2.2mph & 2.9mph (3.5-4.6km/h)
Fuel consumption: 9.64hp/hr per gallon
Weight: 6,819lb (3,100kg)

This tractor was the last design produced by Minneapolis Steel & Machinery as an independent concern – it was introduced for 1929, the year of the merger. The 21-32 was really an enlarged version of the 12-20/17-28, slotting in between that and the bigger 20-35. The basic format of the 12-20 had been around for nearly ten years, but that was sufficiently advanced for the new machine not only to appear modern when it was launched, but to stay in production with only minor changes for 15 years.

The four-cylinder engine was only slightly larger than that of the 17-28, but according to Nebraska it had significantly more power. Rated 21hp at the drawbar, it delivered 31! At first, it had only a two-speed transmission, though a three-speed was later added. A catalogue of the time emphasised the 21-32's

Above: Four valves per cylinder, unit construction – more forward thinking.

per horsepower for more years. The 17-28 has the reputation for keeping lubrication, fuel and upkeep costs down all during its L-O-N-G LIFE. It has that FAMOUS 4-cylinder low speed Twin City engine...BURNS kerosene, gasoline or engine distillate without water injection.'

Above: The final Minneapolis Twin City. Post-merger, M-M was born.

lubrication system, which included a gear-driven pump for the engine, an oil pressure gauge, large reservoir of oil for the gearbox (which also enclosed the steering gear).

Also known as the FT, the 21-32 was renamed FTA in 1935, which signified a slightly bigger engine with a bore of 4.675in (117mm).

Minneapolis-Moline MT

USA

Specifications for Model MT (1931)
Engine: Water-cooled, four-cylinder
Bore x stroke: 4.25in x 5.0in (106 x 125mm)
Capacity: 284ci (4,430cc)
PTO power: 26.7hp @ 1,000rpm
Drawbar power: 18.2hp
Transmission: Three-speed
Speeds: 2.1-4.2mph (3.4-6.7km/h)
Fuel consumption: 5.64hp/hr per gallon
Weight: 5,235lb (2,380kg)

Nineteen-twenty-nine was the year of the merger which created the Minneapolis-Moline Power Implement Company. It was just as well, as it's unlikely that any of those three companies would have survived the Great Depression single-handed. It was a time when tractor and implement ranges were having to expand in response to demand from farmers. For many small concerns, this meant over-reaching themselves, and the only sensible course was to merge with someone making complementary products.

On the face of it, this was the case with Minneapolis-Moline: Moline concentrated on ploughs and the two Minneapolis companies on tractors and threshers respectively. In reality, there was considerable overlap in tractor production. This made economic nonsense, and M-M lost no time in dropping the Minneapolis Threshing's 27-44, as the Minneapolis Steel's 27-42 was superior. The few years after the merger also saw a general renewal of the tractor range: the 17-28 and 27-44 were phased out in 1935, while the KT, FT and MT were all updated the same year, into the KTA, FTA and MTA.

The MTA pictured here, basically a row-crop version of the KTA, shows the layout that had now become the norm for most American row-crop tractors: tricycle wheel format, big four-cylinder engine and transmission forming the 'chassis', fuel tank mounted on top and driver sitting between the rear wheels. It also shows the combination of 'MM' and 'Twin City' logos – the latter was on the way out, while that big chunky 'MM' would soon be the new corporate image.

Below: New 'M-M' logo soon superseded the Twin City.

Minneapolis-Moline Model Z USA

Specifications for Model Z (1950)
Engine: Water-cooled, four-cylinder
Capacity: 185ci (2,886cc)
PTO power: 34.8hp @ 1,500rpm
Drawbar power: 25.1hp
Transmission: Five-speed
Speeds: 2.4mph-13.1mph (3.8-21km/h)
Fuel consumption: 8.55hp/hr per gallon
Weight: 4,290lb (1,950kg)

Although M-M later neglected the small tractors, it did make one in the 1930s. The Universal J of 1934 weighed only 3,450lb and boasted power lift, as well as power take-off. Even bigger news was its five-speed gearbox – a first for tractors – which provided four field-speed gears plus a higher ratio for the road. With rubber tyres, this enabled it to reach 18mph on tarmac. A two-plough machine, it also had adjustable rear treads (54-76in, 1,350-1,900mm) and a

three-fuel four-cylinder engine, 'built to stand the gaff', as an advert of the time put it. A fine tractor, but slightly under powered at 14/22hp

Two years later (or three, depending on whom you believe) M-M replaced it with the Z, which used a slightly smaller all-new 185ci (2.9 litres) four-cylinder motor, albeit with more power. Like the J, the Z had a five-speed transmission and was also designed with easy home maintenance in mind. For example, the valves were mounted horizontally, opened by long vertical rocker arms, easing valve adjustment. Look at a Z from overhead, and the Visonlined styling was clear: the bonnet tapered back towards the driver, to optimise his view of front-mounted implements. And the old grey colour scheme was replaced by Prairie Gold yellow with red wheels. A standard tread version – ZTS – soon followed. Incidentally, M-M's three-letter model names are quite easy to decipher: the first letter (K, J, Z) refers to the model; 'T' stands for tractor; and the final letter refers to chassis type.

Below: M-M's five-speed transmission allowed 18mph on the road.

Minneapolis-Moline Model UTS

USA

Specifications for Model U (1954)
Engine: Water-cooled, four-cylinder
Bore x stroke: 4.25in x 5.0in
Capacity: 283ci (4,415cc)
PTO power: 36.1hp @ 1,300rpm
Torque: 247lb ft @ 949rpm
Drawbar power: 26.8hp
Transmission: Five-speed
Speeds: 2.5-14.0mph (4.0-22.4km/h)
Fuel consumption: 8.66hp/hr per gallon
Weight: 5,905lb (2,684kg)

With the little Z in production, M-M soon followed up with a tractor on the same lines, with similar styling, but much more power. With 39hp, the U was 65 per cent more powerful than the Z, really two classes up. That also made the UTU row-crop version the largest such machine available when it was launched in 1938. It weighed 5,250lb (2,386kg) and was capable of pulling a four-bottom plough. There was a standard-tread version as well, the UTS (pictured here), but both used the same 283ci (4.4 litres) four-cylinder engine with 4.25in bore and five-inch stroke, built in-house by Minneapolis-Moline.

The UTS could be had with steel wheels or pnuematic rubber tyres, though as with its rivals few could be ordered with pneumatics during World War II, due to a rubber shortage. As with other tractors (and especially M-Ms) the U series was a three-fuel tractor, capable of running on gasoline, kerosene or distillate. The latter gave less power than gasoline – 7hp less, on the U series – but was cheaper to buy, and avoided the pre-ignition problems of kerosene. But the power loss, plus the fact that over time distillate would dilute the oil and thus shorten engine life, led many farmers to invest in gasoline. By the time diesel became mainstream in the 1950s, distillate hardly featured.

Right: Model U was a big four-plough machine, with M-M's own 4.4 litre engine.

Minneapolis-Moline UDLX USA

Engine: Water-cooled, four-cylinder
Bore x stroke: 4.25in x 5.0in (106 x 125mm)
Capacity: 283ci (4,415cc)
PTO power: 38hp @ 1,275rpm
Drawbar power: 31hp
Transmission: Five-speed
Weight: 6,000lb (2,727kg)

Today, we take for granted that almost every new tractor will come with an enclosed cab with a radio, heater and windscreen wipers. But this was unheard of in 1938 when M-M announced the UDLX 'Comfortractor' to an astonished world. You didn't clamber up to the UDLX's cab, you stepped into it via a convenient rear door. It had a heater and radio when many cars didn't, and even a passenger seat. There was safety glass, a defroster and foot accelerator.

M-M's trademark high top gear gave a road speed of 40mph (65km/h). It looked elegant too, clean and streamlined in a way that some cars of 1938 failed to emulate. This was a new sort of tractor. There was even a sleek looking cabless variant, which retained the bodywork. In theory, you could plough in comfort all day (whatever the weather), go home, then drive out to town for the evening – all in the comfort of your own tractor.

In practice, the enclosed cab gave less than perfect visibility (especially in the wet – take another look at the size of those wipers!) though the tractor was good enough for field work. And although based on standard U

Minneapolis-Moline UTC USA

Specifications of Model U quoted, as the two were mechanically identical
Engine: Water-cooled, four-cylinder
Bore x stroke: 4.25in x 5.0in (106 x 125mm)
Capacity: 283ci (4,415cc)
PTO power: 36.1hp @ 1,300rpm
Torque: 247lb ft @ 949rpm
Drawbar power: 26.8hp
Transmission: Five-speed
Speeds: 2.5-14.0mph (4.0-22.4km/h)
Fuel consumption: 8.66hp/hr per gallon
Weight: 5,905lb (2,684kg)

This was the third variation of the basic U series theme, the UTC Cane tractor, specifically designed for cane cultivation. Introduced in 1948, it had exceptionally high clearance and an arched front axle, and was available through to 1954. In that year, it cost $3,200, ready to work, and came with the five-speed transmission that was fast becoming an M-M trademark. Even allowing for the extra metalwork in the UTC's high clearance chassis, that reflected a big increase on six years before, when the UTU cost just $1,586 on steel wheels, with rubber a $200 option.

Like the other Us, the Cane was available with LPG power, which Minneapolis-Moline had pioneered in 1941. Like distillate, it was cheaper than gasoline, gave better economy and needed less maintenance. But although M-M was first with LPG, it was tardy in making diesel available. It wasn't until the

Above: Too much, too soon – the Comfortractor.

series components under its shapely skin, the UDLX cost a great deal more, at $2,155. Few farmers could justify that sort of money, and only around 150 Comfortractors were made.

Above: M-M pioneered LPG power, but was late with diesel.

mid 1950s that a diesel powered UB was on sale. This used M-M's own four-cylinder diesel, based on the same 283ci (4.4 litres) block as the other models, and was named the UB Special Diesel.

Minneapolis-Moline R

USA

Specifications for Model R (1951)
Engine: Water-cooled, four-cylinder
Bore x stroke: 3.63in x 4.0in (91 x 100mm)
Capacity: 165ci (2,574cc)
PTO power: 25.9hp @ 1,500rpm
Drawbar power: 18.3hp
Transmission: Four-speed
Speeds: 2.6-13.2mph (4.2-21km/h)
Fuel consumption: 7.95hp/hr per gallon
Weight: 3,414lb (1,552kg)

The Comfortractor may not have been an overwhelming success (though it probably justified itself simply in bringing publicity to M-M) but its maker hadn't abandoned the idea of a steel cab, and carried on offering this on some models. The little R, for instance, a new two-plough tractor designed to fit into the range beneath the 2-3-plough Z. It was really a smaller version of the Z, with a 165ci

engine with all the same easy-maintenance features. Those horizontal valves, for instance, which meant very easy clearance checking. Crankshaft and con-rods could be examined without even draining the sump, and the engine could be overhauled without dismantling the entire machine.
Unveiled in 1939, with 18/25hp from its little four-cylinder engine, the basic R started at $1,500, though there was soon a whole range to choose from. RTU was the row-crop version, with tricycle layout; RTS had a standard-tread (twin front wheels) and RTE had wide-front adjustable front tread. There was also an RTI industrial version. Together they ran through to 1955, though the less successful RTE was dropped a couple of years earlier.

'For the economy-minded farmer a Minneapolis-Moline R is RIGHT'said the ads.' He knows that it is better to have a little extra power than not quite enough. The R offers four kinds of power for immediate use, with economy of cost: drawbar, Uni-Matic Power, power take-off and belt pulley.'

Below: Not in the Comfortractor league, but the R offered a steel cab option.

Minneapolis-Moline GTA

USA

Specifications for Model GTA (1950)
Engine: Water-cooled, four-cylinder
Bore x stroke: 4.63in x 6.0in (116 x 150mm)
Capacity: 403ci (6,287cc)
PTO power: 55.9hp @ 1,100rpm
Drawbar power: 39.2hp
Transmission: Five-speed
Speeds: 2.5-13.8mph (4.0-22km/h)
Fuel consumption: 9.72hp/hr per gallon
Weight: 7,230lb (3,286kg)

M-M might have left behind the days of monster tractors, but in the late 1930s it could see the way things were going. Once again, farmers were clamouring for bigger, more powerful machines, and the company obliged with the GT in 1938. There wasn't much radically new about the GT, and in fact it used the same engine that had originally seen light of day in the 21/32 of 1926. Still, in 403ci form it produced 36hp at the drawbar, 49 at the belt, which made it strong enough to draw a five-bottom plough.

Even more power followed in 1942, when the uprated GTA was announced, with 56 PTO horsepower and 39 for towing duties. The GTA incidentally, can be discerned from the GT by its Prairie Yellow grille, to match the rest of the machine – the GT had a red grille. A GTB followed in 1947, while the LPG-fuelled GTC was available between 1951 and 1953. Finally, the GT just took part in the diesel era as well, with a big 425ci (6.6 litres) diesel available in the GTB

Minneapolis-Moline ZTX

USA

(Mechanically similar to ZT)
Engine: Water-cooled, four-cylinder
Bore x stroke: 3.625in x 4.5in
Capacity: 186ci (2,901cc)
PTO power: 25.2hp
Drawbar power: 19.8hp
Transmission: Five-speed
Speeds: 2.2mph-14.3mph (3.5-22.9km/h)
Weight: 4,280lb (1,945kg)

The Z series remained in production right through World War II: the ZTU and ZTN ran from 1940 to 1948, alongside the ZTS (which was dropped a year earlier). The ZA series (ZAU, S, E and N) took over in 1949. But as with the other tractor lines, there were few updates in wartime – Minneapolis-Moline was heavily committed to producing military vehicles. It also produced a prototype Jeep for the US Navy in 1944. This was very different from the well-known Willys-Overland Jeep, betraying its tractor origins with a bonnet that extended backwards between driver and passenger.

But M-M was working on the conversion of tractors to military usage as early as 1938, and the ZTX pictured here was one of these. It used the engine and chassis of the ZTS, with the cab from the smaller R series machine. The five-speed gearbox gave a top speed of 15.3mph (24.5km/h), and the heavy front protective grille was a feature unique to this model. Oddly, it was also painted the standard Prairie Yellow colour. Only 25 ZTXs were built, and would probably have been used for short haulage of heavy loads. In the early 1950s,

Above: Power generation – 39 drawbar hp made the GTA a five-plough tractor.

chassis from 1953. This GTB-D was produced for two years before being updated as the GBD, giving 44/56hp. But gasoline or diesel, six cylinders were the coming thing for engines of this size, and the GT had only four.

Above: ZTX was a military specification ZTS – only 25 were built.

a military version of the RTI industrial tractor was produced, for use in the Korean War. It was equipped with lifting lugs for loading by crane.

Minneapolis-Moline V

USA

No specifications available

Through the late 1940s, M-M watched as the demand for small tractors increased. It had nothing to offer, since its smallest machine was the two-plough R, but even that was too heavy and expensive for some vegetable or small acreage farmers. The term (and perhaps the practice) of 'hobby farmers' was yet to come. The company had actually built a two-cylinder prototype machine, the YT, based around an R series engine cut in half. Twenty-five of these were built, but all were eventually returned to the factory.

But in 1950 came an opportunity to put this right, and expand the M-M dealer network at the same time. The B. F. Avery Company, a maker of light tractors and equipment in Louisville, Kentucky, came up for sale – M-M bought

t. Of course, it had one eye on Avery's network of dealers across the south-east USA, an area where M-M itself was weak. But it also brought two proven, well established small tractors into the line-up. The single-plough V pictured here) weighed just 1,612lb (733kg) and was powered by a 65ci 1.0 litre) Hercules engine – it was already popular with vegetable and tobacco farmers. And the light two-plough BG (132ci/2.1-litre engine, 2,880lb/1,309kg) allowed M-M to move its own R up as a heavy two-plough machine. Sadly, the little V tractor remained in the M-M line-up for only two years. Never again would the company attempt to crack the small tractor market.

Below: The M-M V was really a rebadged Avery, a one-plough machine.

Minneapolis-Moline 335

USA

Specifications for 335 (1957)
Engine: Water-cooled, four-cylinder
Bore x stroke: 3.63in x 4.0in (91 x 100mm)
Capacity: 165ci (2,574cc)
PTO power: 31.8hp @ 1,600rpm
Drawbar power: 24.1hp
Transmission: Five-speed
Speeds: 2.7mph-15.1mph (4.3-24.1km/h)
Fuel consumption: 9.99hp/hr per gallon
Weight: 3,707lb (1,685kg)

In the early 1950s, M-M looked seriously in danger of falling behind. The two small Avery tractors hadn't lasted long, and much time had been spent on developing the three-wheel Uni-Tractor. The latter was intended to be a cut-price piece of powered equipment, which could free up mainstream tractors at busy times such as harvest. Unfortunately, the Uni-Tractor

needed its own range of specific implements, which made it less economically viable. In the meantime, M-M's mainstream tractor line-up looked increasingly tired.

That changed in 1955 with the launch of the small-medium 335 and 445. Both were modern and up to the minute, bristling with the latest features. Ampli-torc for example, was an option, a two-speed planetary gearset between clutch and gearbox. This effectively doubled the choice of ratios available (to ten forward and two reverse) and being hydraulically operated, could shift between ranges without stopping the tractor or even declutching. M-M's own version of the hydraulic three-point hitch was available as well, as was a live PTO.

The 335 came as gasoline only, with a 165ci (2.6 litres) four-cylinder engine of 24/30hp, and in utility four-wheel form with adjustable front tread. Its bigger 445 brother was powered by a 206ci (3.2 litres) four, available in gasoline, LPG or diesel form, and with 31/38hp.

Below: Ampli-torc doubled the new 335's gear ratios to 10 forward speeds.

Minneapolis Moline M-5 USA

Specifications for Moline M-5 diesel (1960)
Engine: Water-cooled, four-cylinder
Bore x stroke: 4.63in x 5.0in (116 x 125mm)
Capacity: 336ci (5,241cc)
PTO power: 58hp @ 1,500rpm
Drawbar power: 51.4hp
Transmission: Five-speed
Speeds: 3.1mph-17.4mph (5.0-27.8km/h)
Fuel consumption: 13.3hp/hr per gallon
Weight: 6,965lb (3,166kg)

The long-running U series had first seen the light of day in 1938, but M-M didn't replace it until 1957, with the 5-Star. Even this used the same 283ci (4.4 litres) engine in gasoline and LPG form, though the diesel was upped in size to 336ci (5.2 litres). While some manufacturers were tending towards six cylinders for smoothness, and/or shorter strokes for higher revving engines, M-M remained

faithful to the long-stroke low revving fours that had served it well since it first came into the tractor business.The 5-Star did bring new styling though, and now placed the driver ahead of the rear axle and astride the transmission.

For 1960, the M-5 replaced the 5-Star, though it persevered with the 336ci (4.4 litres) diesel and its five-inch (125mm) stroke. It did have Ampli-Torc now, giving ten forward speeds, plus the live hydraulics which all customers expected. Power steering was standard too. The M-5 was joined in 1962 by a four-wheel-drive version, the M504, with mechanical front wheel drive, which M-M said was the industry's first sub-100hp 4x4. The largest M-M was now the Gvi, with a higher seating position and more power than the M-5, while the smaller 335 had been replaced by the Jet Star and the 445 by the 4-Star. What the company lacked was a sub-30hp machine, which put it at a disadvantage compared to rivals like Ford or International.

Below: Only four cylinders, but the M-5 had power steering and Ampli-torc.

M-M Massey-Ferguson 97 USA

Specifications for 97 diesel (using G705 basis)
Engine: Water-cooled, six-cylinder
Bore x stroke: 4.63in x 5.0in (116 x 125mm)
Capacity: 504ci (7,682cc)
PTO power: 101hp @ 1,600rpm
Drawbar power: 90.8hp
Transmission: Five-speed
Speeds: 3.3mph-18.3mph (5.3-29.3km/h)
Fuel consumption: 12.35hp/hr per gallon
Weight: 8,155lb (3,706kg)

No, it's no mistake, the Massey-Ferguson 97 really was built by Minneapolis-Moline. M-M had been here before of course. Minneapolis Steel & Machinery was actually kick-started into the tractor business by building machines under contract for JI Case and the Bull company – these valuable contracts actually helped the company keep afloat while its own

slower selling Twin City giant machines were established.

The M-F 97 was slightly different. Really it was a convenient means for Massey-Ferguson to sell a large tractor for the USA market. So the 97 was built on existing M-M chassis – the G705, 706, 707 or 708. The 705 and 707 came first, the latter being a front-wheel-assist version of the same machine. The 706 and 708 were simply updates, two-wheel-drive and front-wheel-assist respectively. The arrangement evidently wasn't a huge success for M-F or Minneapolis-Moline, as only 429 of these 97s were built. In fact, the M-F 97 wasn't the first co-operation between the two firms – from 1958 M-F had been buying GBD and Gvi machines, changing some sheet metal, the colour scheme and badges, and calling the result the Massey-Ferguson 95. As tractor companies merged or sought new markets by co-operation, this blend of badge-engineering became increasingly common.

Below: Despite the badge, this was a thinly disguised M-M G705.

Minneapolis-Moline G705 USA

Specifications for G-705 Diesel (1963)
Engine: Water-cooled, six-cylinder
Bore x stroke: 4.63in x 5.0in (116 x 125mm)
Capacity: 504ci (7,682cc)
PTO power: 101hp @ 1,600rpm
Drawbar power: 90.8hp
Transmission: Five-speed
Speeds: 3.3mph-18.3mph (5.3-29.3km/h)
Fuel consumption: 12.35hp/hr per gallon
Weight: 8,155lb (3,706kg)

Just as it was determinedly modernising its mid-range tractors in the late 1950s, M-M lost no time in doing the same for its top of the range machines. In 1963, the big Gvi was replaced by the first of the G series, the G705. Significantly, there was no straight gasoline options the choices being diesel or LPG, both based on a

massive 504ci six-cylinder engine which M-M built itself. There was a 706 as well, with four-wheel-drive, but otherwise mechanically the same as the 705.

Originally, both LPG and diesel were rated by the company at 105hp at the flywheel, but they later revised this to 112hp. Nebrasakan tests, however, suggested 102hp, for the diesel at least. Four-wheel-drive was an expensive option, adding $2,000 to the G705 diesel's basic price of $6,725, but it was a sign of things to come – as power inexorably increased, four-wheel-drive became the only practical means of applying it without losing traction. Both 705 and 706 used the latest squared-off but slightly rounded styling, and were finished in yellow with white highlights. Produced until 1965 (though some dealer guides suggest they were on sale until '67) the 705 and 706 were superseded by the 707 and 708. The same new styling also appeared on the M602 and 604, which replaced the M5/504 in 1963, and the smaller U302 of '64.

Below: M-M's own 7.7 litre diesel powered the G705.

Minneapolis-Moline G706

USA

Specifications for G706 LPG (1963)
Engine: Water-cooled, six-cylinder
Bore x stroke: 4.63in x 5.0in (116 x 125mm)
Capacity: 504ci (7,682cc)
PTO power: 101hp @ 1,600rpm
Drawbar power: 86.2hp
Transmission: Five-speed
Speeds: 3.3mph-18.3mph (5.3-29.3km/h)
Fuel consumption: 7.82hp/hr per gallon
Weight: 9,165lb/4,166kg (diesel model)

The M-M G706, as we've seen, was no more nor less than a four-wheel-drive version of the G705. Announced alongside the 705 in 1963, it too had a the choice of LPG or diesel power. But although it had four-wheel-drive the 706 was a conventional tractor in layout. There was a growing market for a far bigger machine, with four-wheel-drive and equal-sized wheels, which had been highlighted by specialists such as Steiger. As the 1960s progressed, the mainstream manufacturers became increasingly interested in this market, either contracting a specialist to build a suitable machine for them, or building one themselves.

The one with the Minneapolis-Moline badge arrived in 1969. The A4T was built in-house, albeit at the White plant at Hopkins, Minnesota and was variously marketed as an Oliver, White or Minneapolis-Moline, depending on which dealer happened to be selling it. This made economic sense, but at the expense of diluting a respected name. It was articulated, a response to the need for ever-larger tractors to retain manoeuvrability, and was powered by the usual 504ci (7.7 litres) six with a 10-speed transmission. The A4T 1400 was on sale 1969-70, and a 1600 version in '70-'72. it was also sold as an Oliver 2655 and White Plainsman.

Right: More power made four-wheel-drive increasingly attractive – G706 was M-M's 705 with the 4x4 option.

MINNEAPOLIS-MOL
G706

Minneapolis-Moline M670 USA

Specifications for M670 petrol (1965)
Engine: Water-cooled, four-cylinder
Bore x stroke: 4.63in x 5.0in (116 x 125mm)
Capacity: 336ci (5,242cc)
PTO power: 73hp @ 1,600rpm
Drawbar power: 62.2hp
Transmission: Ten-speed
Speeds: 1.6mph-17.2mph (2.6-27.5km/h)
Fuel consumption: 9.89hp/hr per gallon
Weight: 7,395lb (3,361kg)

This was the last of the M-series, which had begun with the M-5 in 1960 but whose roots went back much further. The M670 still used Minneapolis-Moline's faithful 336ci (5.2 litres) water-cooled four-cylinder engine, in gasoline, LPG or diesel form. This motor had been used in the M670's predecessor, the M-5, and

the Five-Star before that, and was basically a bored-out version of 283ci (4.4 litres) engine used by the U series tractors from 1938! Supplemented by the four-wheel-drive M504, the M-5 had been superseded by the short-lived M602 (plus an M604 four-wheel-drive variant) which were offered in 1963-64. The M670 lasted longer, being launched in '64 and featuring in the Minneapolis-Moline line-up right through to 1970.

Alongside the M670, and announced the same year, was the smaller U302, which shared styling with the big G705/706. It was a four-bottom plough machine, and used a 221ci (3.4 litres) four-cylinder engine. As launched, it came in gasoline, LPG or diesel versions. All of them were offered until 1972. The U302 featured hydrostatic steering, and a device which ensured that the engine would start only in neutral.

Below: M670 was M-M's final mid-range tractor, post-White takeover.

Minneapolis-Moline G1000/1050

USA

Specifications for G-1000 Vista LPG (1968)
Engine: Water-cooled, six-cylinder
Bore x stroke: 4.63in x 5.0in (116 x 125mm)
Capacity: 504ci (7,682cc)
PTO power: 111hp @ 1,800rpm
Drawbar power: 96.6hp
Transmission: Ten-speed
Speeds: 2.3mph-18.0mph (3.7-28.9km/h)
Fuel consumption: 8.33hp/hr per gallon
Weight: 11,960lb (5,436kg)

Despite the White takeover of 1963, in which M-M had joined with Oliver and Cockshutt to form White Farm Equipment, the company continued to produce

new models, albeit with a strong contribution from existing parts bins. The G1050 pictured here started out as the G1000 in 1965, a new big row-crop tractor which used the now familiar 504ci (7.7 litres) six-cylinder engine in diesel or LPG forms. As with the rest of the range, transmission was five-speed gearbox plus two-speed Ampli-Torc. After a couple of years the driver's station was moved higher and forward, and the fuel tank was moved to behind the seat. This was the G1000 Vista.

There were variations on the same theme. A G900 appeared in 1967, the same chassis but with a 425ci (6.6 litres) engine in diesel, LPG or (increasingly unusual) gasoline forms. All were available as wheatland tractors in two-wheel-drive form. New sheet metal in 1969 heralded the G1050 and G955.

Below: M-M's biggest and last model, the G1050 – LPG model shown here.

Moline Universal

USA

Specifications for Model D (1920)
Engine: Water-cooled, four-cylinder
Bore x stroke: 3.5in x 5.0in (88 x 125mm)
Capacity: 192ci (2,995cc)
PTO power: 27.5hp @ 1,800rpm
Drawbar power: 17.4hp
Transmission: Single-speed
Speed: 3.6mph (5.7km/h)
Fuel consumption: 8.15hp/hr per gallon
Weight: 3,590lb (1,632kg)

The Moline Plow Company was well known for making exactly what its name suggested. But as the tractor market in America began to take off, it could not resist bidding for a slice of the action. A prototype motorised plough was built in 1913, but soon scrapped.

Moline's second attempt was more successful, if a little out of the ordinary. By buying the Universal Tractor Company in 1915, Moline acquired an unusual machine that could claim to be the one of the first row-crop tractors. Instead of having two driven rear wheels and small closely-space fronts, the Universal took the opposite layout. Two big front wheels did the driving, with the engine mounted between them, and two smaller wheels at the rear. The advantages were many. With the motor mounted right over the driving wheels (Moline claimed that 98 per cent of the weight was there) traction was good. Better still, the design allowed for the relatively easy adaptation of existing horsedrawn implements – there was no need to buy new tools. The Moline Universal appeared initially in two-cylinder form, then as a four, but it did not survive beyond 1923. Why? Well, apart from the post-war farming depression, it was the old story – you could buy a Fordson for a fraction of the price.

Below: Back to front? The front-wheel-drive Moline Universal.

New Holland 7740

USA/UK

Engine: Water-cooled, four-cylinder turbo-diesel
Capacity: 321ci (5.0 litres)
Power: 95hp
Transmission: 12-speed (F and R)

It's odd that New Holland has become the badge on so many tractors, and lives on today as part of CNH Global, the result of the merger between Case and Fiat-Ford-New Holland. Odd because New Holland itself never built tractors at all, but implements and (later) combine harvesters.

It was originally founded in Pennsylvania in 1895, and specialised in non-powered implements. A change of ownership came in 1940 (after which it produced one of the first successful automatic pick-up hay balers) and again in

1947, when the Sperry Corporation took over. In 1964, Sperry took a major interest in Claas, then one of the largest combine manufacturers in Europe. Sperry New Holland concentrated on combines and implements until 1986, when it was taken over by Ford.

The New Holland name was evidently prized, since in the late 1990s, even well after the Fiat takeover, all tractors apart from the range-topping 70 series were badged as New Hollands, though they wore Ford's blue/black colour scheme. The 7740 pictured here was part of the mid-range 40 series, with a turbocharged four-cylinder diesel producing 95hp.

Below: New Holland became a tractor marque almost by default. The name appeared on tractors only after the Ford takepver.

New Holland 7840

USA/UK

Engine: Water-cooled, four-cylinder turbo-diesel
Capacity: 321ci (5.0 litres)
Transmission: 12-speed (F and R)

New Holland's 40 series was highly successful, and the 75,000th example was built at the group's plant in Basildon, England, in 1996. Also from that year, it was sold as the Fiat S series in some European countries, just as the 70 series was badged as a Fiat G series for some markets.

In fact, '96 was a busy year for the Fiat/New Holland group (now named NH Geotech). A new smaller tractor, the 35 series, was announced. This was Italian-built, and covered the 65-95hp range – as Fiat, it was the L series, as a New Holland, the 35. As if that wasn't enough, 1996 also saw the launch of the

60 series 100-160hp range, this time built at Basildon. Once again, it came with either Fiat or New Holland badges, and was sold as the 60 or M-series. It came with the modern low-nose look that many tractors adopted in the late 1990s, notably the dramatic looking Deutz Agrotron. All the 60 series were powered by a six-cylinder 7.6 litre diesel which was capable of meeting the latest EPA emissions regulations – pollution control was now extending from road vehicles into the field.

As for the 7840 shown here, it was the best-selling model of the 40 series – in fact, in Britain it was the best-selling tractor of any type. But even best-sellers get replaced eventually, and in this case that came in early 1999 with the TS115.

Below: Look closely, there's still a Ford badge on this N-H.

New Holland 8340

USA/UK

Specifications for 8670A (2002)
Engine: Water-cooled, six-cylinder turbo-diesel
Capacity: 458ci (7.5 litres)
Power: 145hp
Transmission: Sixteen-speed Powershift
Fuel economy: 20hp/hr per gall
Weight: 17,250lb (7,824kg)

This tractor, the 8340, was the top of the 40 series range in the mid-1990s. Unlike the smaller four-cylinder 40s, this one used a turbocharged six of 7.5 litres – power was 125hp. The standard transmission was a 12-speed (forward

and reverse) syncromesh unit, with the optional dual-range powershift mulitplying that to 24x24. Or there was the 16-speed Electroshift, which was electronically controlled and itself could be upgraded to 24-speed via a set of creeper ratios.

But popular and adaptable though it was, the 40 series would sooner or later have to be replaced. It happened in 1998, with the new TS range. There were three of them, all with a choice of two- or four-wheel-drive, and all with the low rounded bonnet line already seen on the smaller 35 and 60 series. The 5.0 litre Powerstar four-cylinder diesel came in 80hp non-turbo and 90 or 100hp

Below: N-H's popular 40-series was topped by the turbo'd 8340.

turbo forms: the TS90, TS100 and TS110 respectively.

The following year, a higher-powered six-cylinder TS115 (though rated at 100hp) was announced to take over from the 7840. It used New Holland's own Powerstar 7.5 litre engine and a 16-speed transmission. A 24-speed 'Dual Command' unit was optional, and four wheel-drive was standard on the TS115.

In the year 2000, NH Geotech merged with Case IH to form CNH Global, which within months was closing plants in the USA, Canada, Britain and Italy. Two years on, the future still looks uncertain.

Below: The 40-series was eventually replaced by the TS range.

Nuffield Universal

UK

Specifications for PM-4
Engine: Water-cooled, four-cylinder
Bore x stroke: 4.74in x 4.72in (119 x 118mm)
Capacity: 208ci (3,244cc)
PTO power: 45.4hp @ 2,000rpm
Torque: 223lb ft @ 1,329rpm
Drawbar power: 32.9hp
Transmission: Five-speed
Speeds: 2.3mph-17.3mph (3.6-27.7km/h)
Fuel consumption: 13.77hp/hr per gallon
Weight: 7,011lb (3,187kg)

The Nuffield tractor arose directly out of Britain's desperate economic state after World War II. Heavily in debt, it needed to increase both food production and exports of manufactured goods. A new British-built tractor would help towards both these aims, and the government of the day encouraged the giant Nuffield group (makers of Morris cars) to design and build a suitable machine. The

resulting Nuffield Universal was ready for production as early as 1946, but it was another two years before it finally went on sale, owing to steel shortages.

The new tractor came in M3 row-crop or M4 utility versions, both at just under £500 – that was the basic price, and you had to pay another £60 for three-point hydraulics and a PTO. The engine was Nuffield's own, initially a four-cylinder side-valve TVO of 38hp at 2,000rpm. The new tractor worked well (especially the optional three-point hitch) but because Nuffield was a new name to farmers, the company styled it to look like a Fordson and painted it Allis-Chalmers orange! New engine options soon followed, with a 38hp Perkins diesel in 1950 – the latter was replaced by Nuffield's own diesel in 1954. Diesel was essential for the UK tractor market, and by 1955, all but 5 per cent of Nuffield tractors were so powered.

Whatever engine it used, the Nuffield Universal was a success – it established Morris in the tractor market, and nearly 80 per cent of Universals were exported.

Below: The Nuffield Universal was built for export, and succeeded.

Oliver

USA

Like many American companies that eventually got into the tractor business, Oliver had been making agricultural machinery for a long time. It was founded in 1855 by Scottish-born James Oliver, who had patented a process for producing a plough steel that was tough, with a very long life. The name of his company – the Oliver Chilled Plow Company – referred to this process. However, it had little tractor expertise, and despite having produced a prototype machine, felt unable to put it into production.

The solution came with the 1929 merger with Hart-Parr, which had no new

prototypes but certainly knew how to build tractors. The Hart-Parr name was soon dropped, leaving Oliver as the sole tractor brand of the new Oliver Farm Equipment Sales Company. Three decades of independence followed, before the company was itself taken over by the White Motor Corporation. Under White, the Oliver name gradually faded out of use.

Below: Oliver's styling came right up to date with the 1800 series - six cylinder petrol or diesel.

Oliver Model A/28-44 USA

Engine : Water-cooled, four-cylinder
Bore x stroke: 4.75in x 6.25in (119 x 156mm)
Capacity: 443ci (6,911cc)
PTO power: 49hp @ 1,125rpm
Drawbar power: 28hp
Transmission: Three-speed
Speeds: 2.7-4.3mph (4.3-6.9km/h)
Weight: 6,415lb (2,916kg)

Oliver tractors were really Hart-Parrs, at least partly. Hart-Parr had been in the tractor business since 1903, but found its big heavy machines increasingly outdated as the 1920s progressed – even by the late '20s it offered no row-crop tractor at all. It was saved by a merger in 1929 with three other 'seasoned but faded,' (according to author P. W. Ertel) agricultural machinery makers: the American Seeding Machine

Company (seeders and planters); Nichols and Shepherd (threshers); and the Oliver Chilled Plow Works. Oliver actually had a new tractor on the drawing board, but needed Hart-Parr's production expertise to make it reality.

The result was the Oliver-Hart-Parr Row Crop of 1930, though the Hart-Parr name was dropped after a few years. This first Model A was rated at 18/27 horsepower, with a single front wheel and the rear pair adjustable by sliding them in and out on a splined axle. It was a simple, mechanical means of altering the rear track to suit different crop widths, and became standard for all row-crop tractors over the following decades. The uprated 22/44hp Model A soon followed, with standard treads, along with a Model B 18/28 standard tread. Both soon became known by their power ratings only, and were available in a variety of agricultural as well as industrial forms.

Below: The Model A's rear wheel tread was adjustable on splined axles.

Oliver 70 Row Crop

USA

Specifications for 70 Row Crop (1940)
Engine: Water-cooled, six-cylinder
Bore x stroke: 3.13in x 4.38in (78 x 110mm)
Capacity: 202ci (3,151cc)
PTO power: 30.4hp @ 1,500rpm
Drawbar power: 22.7hp
Transmission: Six-speed
Speeds: 2.6-13.4mph (4.1-21.4km/h)
Fuel consumption: 9.81hp/hr per gallon
Weight: 4,370lb (1,986kg)

Apart from their splined rear axle, the first Olivers were fairly conventional machines. But the Oliver Model 70 announced in 1935 was a true step forward. Until then, most tractors had used low compression, low revving gasoline or distillate engines, usually of two or four cylinders – the diesel era hadn't yet arrived. What made the Oliver 70 different was its high compression six-cylinder motor, built by the Waukesha Motor Company, which incidentally built all of Oliver's tractor engines. Not only that, but stream-lined, car-like styling from 1937 – the 'Fleetline' – set the 70 apart from its boxy contemporaries.

This particular one was small for a six, at 202 cubic inches (3.2 litres), but that made it particularly smooth compared to the less sophisticated opposition, and more powerful of course. 'Power when you want it,' trumpeted an early advertisment, 'power when you need it – power when field conditions are right and time and help are limited. That's the power you'll find in an Oliver Row Crop

Above: Oliver 70 was slim, stylish and smooth.

70.' The only downside was that the 70 needed expensive high octane fuel – for farmers who really couldn't bring themselves to pay out for such a rich diet, the 70 was also available in low-compression distillate form as well.

Below: The 70's high compression required high octane fuel.

Oliver 70 Orchard

USA

Specifications for 70 Orchard (1940)
Engine: Water-cooled, six-cylinder
Bore x stroke: 3.13in x 4.38in (78 x 110mm)
Capacity: 202ci (3,151cc)
PTO power: 30.4hp @ 1,500rpm
Drawbar power: 22.7hp
Transmission: Six-speed
Speeds: 2.6-13.4mph (4.1-21.4km/h)
Fuel consumption: 9.81hp/hr per gallon
Weight: 4,370lb (1,986kg)

The six-cylinder 70 was a real success for Oliver, with nearly 63,000 of them built between the mid-1930s and 1948. Not all of those were standard row-crop machines, and just like earlier Olivers the 70 came in a variety of forms, such as the Orchard version pictured here. One specialised version was the Airport 25, designed for towing aircraft. The green colour was said to have been the result of extensive market research, 1930s style. Several 70s were painted in a variety of colours – red, green, orange, silver and gold – and displayed at country fairs to gauge public reaction. Green got the most favourable reviews, so that's how Olivers were finished from then on.

Customers may not have had a colour choice, but there were lots of optional extras on the 70. As well as the usual rubber tyres (in place of the standard 'tip-toe' steel rims) there was a popular electric start option, a wide front axle, mudguards and, if you stuck with the tip-toe rims, road bands –

Oliver 80

USA

Specifications for 80 Standard HC (1940)
Engine: Water-cooled, four-cylinder
Bore x stroke: 4.25in x 5.25in (106 x 131mm)
Capacity: 298ci (4,649cc)
PTO power: 38hp @ 1,200rpm
Drawbar power: 28hp
Transmission: Four-speed
Speeds: 2.8-6.4mph (4.5-10.2km/h)
Fuel consumption: 9.5hp/hr per gallon
Weight: 8,145lb (3,702kg)

Oliver might have made a bold step forward with the six-cylinder 70, but it wasn't about to drop the proven four-cylinder Hart-Parrs. The Oliver 80, announced in 1937 was based on the old Hart-Parr 18-27 and 18-28, with the Oliver-Hart-Parr 80 industrial tractor the link between them.

On sale for the 1938 model year, there was little new about the Oliver 80, still powered by a simple four-cylinder motor in high-compression gasoline or low-comp kerosene versions. Power outputs for the two were virtually the same, though the kerosene was slightly bigger to counteract its lower efficicnecy – bore was 4.50in against 4.25in. A diesel option was added for 1940, using a Buda-Lanova engine, and the year before. Oliver had updated the 80 with ASAE standard PTO and hitches, plus a four-speed gearbox.

Right: If the six-cylinder 70 was too racy, Oliver still offered the slower-revving four-cylinder 80, which owed much to the older Hart-Parrs.

together these could push the price of an Oliver 70 to over $1,000. Still, with the 70's handsome streamlined appearance, you would end up with one of the best looking tractors for miles around!

Right: The 70 Orchard's graceful fenders were there for a purpose.

Oliver 90

USA

Specifications for Oliver 90 (1937)
Engine: Water-cooled, four-cylinder
Bore x stroke: 4.75in x 6.25in (119 x 156mm)
Capacity: 443ci (6,911cc)
PTO power: 49hp @ 1,125rpm
Drawbar power: 28hp
Transmission: Three-speed
Speeds: 2.7-4.3mph (4.3-6.9km/h)
Weight: 6,415lb (2,916kg)

Just as the Oliver 80 was no more than a rebadged (and later updated) version of the original Hart-Parrs, so was the 90. This was an update on the biggest Hart-Parr, the 28-44 or Model A, which by 1937, when the 90 was announced, had already been in production for a few years. However, it was more comprehensively upgraded than the 80. Oliver used the same big 443ci (6.9 litres) four-cylinder engine, rated at 1,125rpm and with 4.75in x 6.25in (119 x 156mm) dimensions. But now it had high-pressure lubrication, an electric start and centrifugal governor. There were both gasoline and kerosene versions, but no high compression option. Transmission was four-speed as standard, but a PTO cost extra.

The 90 was a long-lived tractor, despite tracing its roots back to the early 1930s, and production wasn't finally wound up until 1953. In its last year, some 90s used six-cylinder engines, while production was moved from Charles City to the Oliver plant in South Bend.

Oliver 60

USA

Specifications for Row Crop 60 (1941)
Engine: Water-cooled, four-cylinder
Bore x stroke: 3.13in x 3.5in (78 x 88mm)
Capacity: 108ci (1,685cc)
PTO power: 18.3hp @ 1,500rpm
Drawbar power: 13.6hp
Transmission: Four-speed
Speeds: 2.6-6.1mph (4.2-9.8km/h)
Fuel consumption: 10.13hp/hr per gallon
Weight: 2,450lb (1,113kg)

Oliver's tractor line-up was filling out nicely by the late 1930s, with the 70 slotting in beneath the bigger ex-Hart-Parr machines. But it still didn't have a small tractor to compete with the Allis-Chalmers B, John Deere H and Farmall A – not to mention the fruits of the Ford-Ferguson partnership.

Not surprisingly, given the success of the six-cylinder 70, when the new 60 arrived it turned out to be a smaller version of the same thing. The engine might have had four cylinders in place of six, but really it was a four-cylinder version of the same motor, running to the relatively high speed of 1,500rpm – and, like the 70, there was a four-speed transmission. Rubber tyres were standard too, though during World War II the 60, like many other tractors, had to temporarily revert to steel wheels, thanks to a rubber shortage. Whatever, the modern engine, rubber tyres and four speeds underlined the fact that this was no cut-price special. An electrical system cost extra though, while ignition varied between Wico magnetos and car-

Above: The 90 was another Oliver with Hart-Parr origins.

Above: Oliver's 60 was a true smaller brother to the 70.

type distributor systems. The 60 established Oliver in the smaller tractor market, and was in production until 1948.

Oliver 66

USA

Specifications for Row Crop 66 (1949)
Engine: Water-cooled, four-cylinder
Bore x stroke: 3.13in x 3.75in (78 x 94mm)
Capacity: 129ci (2,012cc)
PTO power: 24hp @ 1,600rpm
Drawbar power: 16.8hp
Transmission: Six-speed
Speeds: 2.5-11.4mph (4.0-18.2km/h)
Fuel consumption: 9.89hp/hr per gallon
Weight: 3,193lb (1,451kg)

Just like its bigger brothers, the little 60 was updated into the double figure series. It made the change in 1947, alongside the 77, though the big 99 had already been unveiled before World War II. The main difference between 60 and 66 centred around power units. The 66 continued with basically the same motor, albeit with a longer stroke of 3.75in (94mm), which yielded a larger capacity of 129ci (2.0 litres). Apart from the basic dimensions, this was still basically a four-cylinder version of the 70's proven Waukesha-built six.

There was a wider choice of engines as well, with a diesel option for the first time, plus the usual kerosene-distillate. In the latter case a slightly larger bore gave 145ci (2.3 litres), to make up for kerosene's lower efficiency. The 66 – still Oliver's baby tractor, stayed in production for six years, until replaced by the Super 66 in 1954, using Oliver's own four-

Oliver 77

USA

Specifications for 77 (1952)
Engine: Water-cooled, six-cylinder
Bore x stroke: 3.31in x 3.75in (83 x 94mm)
Capacity: 194ci (3,026cc)
PTO power: 34.5hp
Drawbar power: 25.8hp
Transmission: Six-speed
Speeds: 2.5-11.6mph (4.0-18.6km/h)
Weight: 4,670lb (2,123kg)

The six-cylinder 70, which had so revolutionised Oliver's line-up back in 1935 (not to mention the mid-range tractor market) was finally replaced by the three-plough 77 a dozen years later. It was still based around a six-cylinder engine, but a juggling with bore and stroke dimensions (larger bore, shorter stroke) brought slightly less capacity, slightly higher engine speeds and a little more power. The gasoline and diesel options now came out at 194ci ((3.0 litres), with 216ci (3.4 litres) for the kerosene-distillate.

The kerosene motor was dropped after just a couple of years, and 1952 saw a liquid petroleum gas (LPG) option introduced, based on the standard gasoline engine. In fact, the late 1940s/early 1950s were a virtual battleground of the various power options. Kerosene/distillate was cheap but inefficient, and on the way out; gasoline gave best power, but was thirsty; LPG promised much of the power of gasoline with a cheaper fuel; and the new diesels cost more to make and buy but were most fuel-efficient of all. Oliver's own Hydra-Lectric hydraulic lift, which provided

Above: Oliver's smallest – the 60 – was updated as the 66 in 1947.

cylinder engine. The Super also added a three-point hitch with Oliver's Hydra-Lectric hydraulic lift, and a three-speed governor. The Super in turn was replaced by the 660 in 1959.

Above: The 77 lacked the elegance of its predecessor.

either electric or manual control of implements, was another update for the 77, in 1949.

Oliver 88

USA

Specifications for Row Crop 88 Diesel (1950)
Engine: Water-cooled, six-cylinder
Bore x stroke: 3.5in x 4.0in (88 x 100mm)
Capacity: 231ci (3,603cc)
PTO power: 43.5hp @ 1,600rpm
Drawbar power: 29.4hp
Transmission: Six-speed
Speeds: 2.5-11.8mph (4.0-18.9km/h)
Fuel consumption: 12.9hp/hr per gallon
Weight: 5,680lb (2,582kg)

Nineteen forty-seven was a busy year for Oliver. To celebrate the name's 100th birthday, the new Fleetline Series of tractors was unveiled – the 66, 77 and 88. Since the original Fleetline 70 of 1937, Olivers had become known for their svelte, streamlined styling – more like a car or truck than a tractor.

To cut costs, there was great interchangeability of parts between the three new machines, though only the 88 offered a diesel option from the outset. It had a six-speed transmission (with two reverse ratios) and a thermostatically controlled cooling system. Rated as a four-plough tractor, the 88 came in three row-crop versions: dual narrow-front, single front wheel and adjustable wide-front. Standard-tread, orchard, industrial and high-crop versions were available as well. As if that wasn't choice enough, most of these were offered with the standard 231ci (3.6 litres) six-cylinder in diesel or gasoline, or in bigger bore 265ci (4.1 litres) kerosene forms. Compression ratios varied from 4.75:1 for the kerosene, to 15.5:1 for diesel, though all motors were rated at 1,600rpm. Like its smaller brothers, the 88 gained the option of Hydra-Lectric hydraulics in 1949. The Super 88 replaced it in 1954.

Below: Oliver's biggest, the 88 came in diesel, gasoline or kerosene forms.

Oliver 99

USA

Specifications for Super 99 GM (1955)
Engine: Water-cooled, three-cylinder two-stroke
Bore x stroke: 4.5in x 5.0in (113 x 125mm)
Capacity: 213ci (3,322cc)
PTO power: 72hp @ 1,675rpm
Torque: 504lb ft @ 1,309rpm
Drawbar power: 59hp
Transmission: Six-speed
Speeds: 2.6-13.8mph (4.2-22.1km/h)
Fuel consumption: 12.4hp/hr per gallon
Weight: 10,155lb (4,616kg)

Oliver's biggest tractor of the day – the 99 – was also one of its most successful, staying in production for 20 years. It had its roots in the 99 Industrial Special High Compression, but a purely agricultural version was first offered for the 1938 model year, replacing the 90. Initially, it used the same 443ci (6.9 litres) four-cylinder engine as the 90, albeit in high-compression form, but six-cylinder 302ci motors soon followed, in both gasoline and diesel forms.

Perhaps more technically interesting was the Super 99 GM diesel. While retaining Oliver's own conventional six-cylinder motors, the Super 99 from 1955 also added the option of a General Motors 3-71 two-stroke. This was a supercharged diesel, with three cylinders, though rated at the same 1,675rpm as the four-stroke sixes. With a 17.0:1 compression ratio, it employed a Roots-

Above: Interesting option – the Super 99 used GM's two-stroke diesel.

type blower to purge exhaust gases and supercharge the fresh mixture. Unlike a conventional two-stroke, this one used poppet exhaust valves. It produced a unique wailing exhaust note, and needed to be kept revving to produce power. Not your average diesel, then. With over 70 horsepower, the 99 GM was the largest, most powerful tractor of its day.

Below: Interesting option – the Super 99 used GM's two-stroke diesel.

Oliver 770

USA

Specifications for 770 diesel (1958)
Engine: Water-cooled, six-cylinder
Bore x stroke: 3.5in x 3.75in (88 x 94mm)
Capacity: 216ci (3,369cc)
PTO power: 48.8hp @ 1,750rpm
Drawbar power: 34.9hp
Transmission: Six-speed
Speeds: 2.1mph-10.8mph (3.4-17.3km/h)
Fuel consumption: 12.9hp/hr per gallon
Weight: 5,565lb (2,530kg)

By the late 1950s, when Oliver's 770 replaced the Super 77, transmission development was proceeding apace. Until then, tractor gearboxes had gradually acquired more ratios – three, four, five and six – but now something new was happening. Oliver's 'Power Booster' (like International Harvester's 'Torque Amplifier') was simply a fancy name for an additional two-speed gearbox that

sat in front of the main six-speeder. When the going got sticky, the driver could get a half-step downshift without loss of power – the actual ratio was 1.32:1. The Power Booster operated on all of the 770's gears, effectively giving it twelve forward speeds and four reverse.

Otherwise, the 770 used the same six-cylinder engines as its predecessor, albeit with a 10 per cent power increase across the range, thanks to higher compression ratios and rated speed upped from 1,600rpm to 1,750rpm. The new Power Traction hitch (shared with the 660) was aimed (as you might expect) at improving traction – the lower links were attached further forward than normal, underneath the chassis, to put more weight on all four wheels, though still with a rearward bias. And finally, like all Oliver 'three-figure' machines, the 770 sported a new colour scheme of 'meadow green' with 'clover white' wheels – who says tractor makers can't get lyrical?

Below: Power Booster was Oliver's version of the two-range transmission.

Oliver 1855

USA

Specifications for 1855 diesel (1970)
Engine: Water-cooled, six-cylinder
Bore x stroke: 3.875in x 4.375in (97 x 109mm)
Capacity: 310ci (4,836cc)
PTO power: 99hp @ 2,400rpm
Drawbar power: 82.7hp
Transmission: Eighteen-speed
Speeds: 1.4-18.6mph (2.2-29.8km/h)
Fuel consumption: 12.2hp/hr per gallon
Weight: 11,140lb (5,063kg)

If Oliver's 1655 was spiritual successor to the original six-cylinder 70, then the 1855 of the same year filled a similar role for the big 80 and 90. Its more recent roots lay in the 1800 series of 1962, the first four-numbered Olivers which

brought styling right up to date. Like many tractors of the period, they had squared-off 'industrial' styling, a far-cry from the streamlined Fleetline look that the 70 had pioneered in 1937. But then, tractors mirrored the general trends of public fashion and design – cars, trucks and refrigerators were all getting the squared off treatment.

The 1800 actually used the same 265ci (4.1 litres) six-cylinder engine as the 880 it replaced, but with rated speed increased to 2,000rpm. The diesel version was slightly larger at 283ci (4.4 litres), and both powered what was a six-plough tractor. Interestingly, Oliver still offered a supercharged General Motors two-stroke diesel, in the higher-powered 1900. In four-cylinder 212ci (3.3 litres) form, that produced 90hp at the drawbar and made the 1900 an eight-plough machine.

Below: Squared-off styling for the 1855 of 1970.

Oliver 1655

USA

Specifications for 1655 (1970)
Engine: Water-cooled, six-cylinder
Bore x stroke: 3.75in x 4.0in (94 x 100mm)
Capacity: 265ci (4,134cc)
PTO power: 70hp @ 2,200rpm
Drawbar power: 57.4hp
Transmission: Eighteen-speed
Weight: 7,780lb (3,536kg)

The 1960s were less certain times for Oliver. In November 1960 it was taken over by well-established truck maker White Corporation. White evidently wanted to get into tractors, as it added the Canadian Cockshutt concern to its stable in 1962, and Minneapolis-Moline the following year. The Cockshutt connection was nothing new for Oliver, which had had a

narketing agreement with the Canadians since the 1930s – Hart-Parr had lone the same even before that. Right through, various Olivers were sold n Canada as Cockshutts, with no more than a different badge and colour cheme. This gave White a good foundation to build on, as it determined to naximise economies of scale by having Oliver, Cockshutt and Moline all hare components.

3y the time this Oliver 1655 was built in 1970, the links were well-established, and the year before White had brought all its tractor technology ogether into a new line under the White Field Boss name. As for the 1655, his was a competitive 70hp machine with three auxiliary transmission options: Hydra-Power Drive, Over/Under Hydraul-Shift or Creeper Drive. And, just like the first 70, this one used a six-cylinder engine.

Below: 1970, and the Oliver badge would soon fade into obscurity.

Oliver 2255

USA

Specifications for 2255 diesel (1973)
Engine: Water-cooled, V8
Bore x stroke: 4.5in x 4.5in (113 x 113mm)
Capacity: 473ci (7,379cc)
PTO power: 147hp @ 2,600rpm
Drawbar power: 124hp
Transmission: Eighteen-speed
Fuel consumption: 13.4hp/hr per gallon
Weight: 16,407lb (7,458kg)

Right through the 1960s and into the '70s, there was a power race amon tractors. Different manufacturers had different solutions. At first, it was enoug to simply increase engine capacity, the number of cylinders, compression ratio and engine speeds. But this soon came up against the law of diminishin

returns: tractor engines need to be long-lived and reliable, with lots of low-rev lugging power. Diesel was in part an answer to that, but was less powerful than gasoline, cubic-inch for cubic-inch. Allis-Chalmers and others pioneered turbocharging diesels, and a turbo-diesel is now the standard big tractor engine. Intercooling would follow in a further quest for more power.

But Oliver didn't go the turbo route (the GM diesels were supercharged, a different priniciple). The 2255 of the early 1970s typified this approach. Instead of a smaller turbocharged six-cylinder diesel, it chose a huge 473ci (7.4 litres) V8 bought in from Caterpillar, the crawler manufacturer. This gave 145hp, not much less than the latest turbo-intercooled Allis-Chalmers six. And it was one of the last powerful two-wheel-drive tractors, before 4x4 proved to be a more efficient means of getting lots of power down to the ground.

Below: End of an era? V8 2255 was one of the last Olivers.

Ransomes MG2

UK

Specifications for Ransomes MG5 (1953)
Engine: Water-cooled,
Power: 5hp
Transmission: Single-speed forward/reverse
Weight: 1,800lb

Ransomes, based in Ipswich, England, built a miniature crawler tractor. The company was making powered lawn mowers in the 1930s, the larger of which were fitted with a 600cc side-valve engine built by Sturmey-Archer, makers of the famous cycle hub gear. This engine, someone evidently realised, was big enough to power a light tractor.

Rather than tackle such a project on their own, Ransomes entered into an agreement with Ford, and with Roadless Traction, the company which of course was a prolific builder of half-track and full-track conversions for Fordson machines. The result, unveiled in 1936, was the Ransomes MG2 machine – MG

stood for Market Garden, as the little tractor was aimed at the legion of small-scale vegetable growers in England – the market gardeners, as they were called. The concept was so successful that Ransomes would be making small crawlers for the next 30 years.

The MG2 used a directly mounted rear frame for implements – single-bottom ploughs and so on – and was arguably a cruder, simpler version of Harry Ferguson's three-point hitch. So popular was the Ransome – 15,000 were made over three decades – that a whole range of implements to suit it were soon on the market, an example being the Demon Spray Pump made by A&G Cooper of Wisbech. The MG2 itself developed as well, into variants such as the MG6 pictured here. When production ended in 1966, the MG40 was diesel powered and equipped with a hydraulic lift for implements.

Below: Ransomes' little crawler was aimed at market gardeners.

Renault Super D

FRANCE

Specifications for GP (1919)
Engine: Water-cooled, four-cylinder
Power: 30hp
Transmission: Three-speed

Renault's first tractor of 1919 was a crawler, actually based on a wartime tank. A wheeled version soon followed, with the same four-cylinder 30hp gasoline engine. In distinctive Renault fashion, it had a curved bonnet, with the radiator mounted behind the engine. Usually, the steering was by tiller, and a wheeled version, the HO, soon followed. Unlike other European companies which built semi-diesel tractors, Renault stuck with gasoline/kerosene spark-ignition engines, and experimented with methane as a fuel during World War II. The

French Army and Academy of Agriculture actually encouraged the development of gas-powered tractors from the 1920s.

After World War II, the 3040 formed the core of Renault's tractor range, and was claimed to be the first machine with a full electrical system. It also had a two-speed PTO, hydraulic lift and adjustable tread. The D series of 1956 was unusual in offering air-cooled as well as water-cooled diesels, and the rest of the tractor – differential lock, syncromesh gearbox and 540rpm PTO – was fully up to date. The exception was the lack of a three-point hitch. For that, buyers had to wait until 1965 and the Super D series pictured here.

Below: Renault's Super D offered diesel power and 3-point hitch.

Renault 180.94

FRANCE

Specifications for Atles 935
Engine: Water-cooled, six-cylinder, turbo-intercooled
Capacity: 462ci (7.2 litres)
Power: 250hp
Transmission: 18-speed F/9-speed R

Renault made a good job of keeping its tractors up to date. It had brought in four-wheel-drive and torque converters in the 1960s, with heaters and anti-vibration cabs adding to driver comfort. It did not build huge super-tractors, concentrating as it did on the European market, but was not shy of high-tech equipment – Renault tractors were some of the first to be fitted with information systems including economy displays.

A complete new range in 1974 covered the 30hp to 115hp range with two-

and four-wheel-drive, plus the innovation of a forward/reverse shuttle transmission, to give quick and easy changes of direction. There were more cab advances in the 1980s: a passenger seat, roof hatch and better visibility in the TX cab of 1981, and the TZ in 1987, which was mounted on springs and shock absorbers to effectively isolate the driver from bumps and and vibration – it won a Gold Medal from RASE.

The 180.94 was the top of Renault's range in the mid-1990s, and used that TZ cab. Its Multishift transmission offered 27 speeds in forward or reverse, thanks to shuttle, and a powershift within nine gears in each of three ranges. By 1999, it had been replaced by the 185hp Ares, but in 2001 that was surpassed by the Atles, Renault's most powerful tractor yet and a move up to the super-tractor class.

Below: Independent, Renault maintained a high-tech approach.

Rock Island/Heider

USA

Specifications for Rock Island 18-35
Engine: Water-cooled, four-cylinder
Bore x stroke: 4.5 x 6.0in (113 x 150mm)
Capacity: 382ci (5,959cc)
PTO power: 36.5hp
Drawbar power: 30.4hp
Transmission: Two-speed
Speeds: 3 and 4.5mph (4.4 and 7.2km/h)
Fuel consumption: 7.1hp/hr per gallon
Weight: 5,740lb (2,574kg)

Heider was based in Cedar Rapids, and had been making tractors from as early as 1910. However, it could not survive on its own, and sold out to implement maker Rock Island Plow Company in 1918. Tractors continued to be badged as

Heiders for several years, but in the late 1920s the Rock Island name was gradually adopted. These Heiders were not revolutionary tractors. The petrol/kerosene Model D of 1919 was rated at 9 drawbar hp, 16hp at the PTO. Transmission was by friction discs, giving a speed range of 1-4mph (1.6-6.4km/h).

The 18-35 pictured here came from the later Rock Island era, and used a Buda four-cylinder engine, Stromberg M3 carburettor and Dixie-Splitdorf-Aero magneto. The G2 of 1929 was really a smaller version of the same thing, rated at 15/25hp. This tractor used a Waukesha four-cylinder engine and weighed 4,200lb (1,890kg).

Finally, in 1937, Rock Island/Heider was taken over by JI Case, which ended its tractor production.

Below: Heider or Rock Island badges were carried, according to era.

Rumely 6A

USA

Specifications for 6A (1930)
Engine: Water-cooled, six-cylinder
Bore x stroke: 4.25in x 4.75in (106 x 119mm)
Capacity: 504ci (7,862cc)
PTO power: 43hp @ 1,365rpm
Drawbar power: 33.6hp
Transmission: Three-speed
Speeds: 2.8-4.7mph (4.5-7.5km/h)
Fuel consumption: 5.71hp/hr per gallon
Weight: 6,370lb (2,867kg)

Advance-Rumely tractors looked seriously outdated by the late 1920s. The massive Oil Pulls, with their great, slow-revving twin-cylinder oil-cooled kerosene engines, looked more like steam traction engines than modern field tractors. In power-to-weight ratios, economy, price and sheer ease of use, they had long since been overtaken by the new breed of gasoline tractors.

Belatedly, the company tried to play catch-up. Smaller versions of the Oil Pull were just too heavy for their power, and in 1927 Rumely bought the Toro Motor Cultivator line – the little four-cylinder DoAll was the result, which weighed a reasonable 3,000lb (1,350kg) and used a conventional Waukesha engine. But it was not a success, and the remaining stocks were sold off to the dealers at a bargain $543 – not much more than half the original list price.

Then at the eleventh hour, just before it was driven into the arms of Allis-Chalmers, Rumely announced another modern tractor, the 6A. This was as far from the old Oil Pulls as it was possible to get: modern Waukesha six-cylinder engine; a six-speed gearbox where the opposition offered three. It could pull 4,273lb (1,923kg) at 9.96 per cent wheel slippage. Maybe this was the tractor that could save Advance-Rumely. Sadly, it didn't, and sales were poor. In fact, 700 unsold 6As were still held at the factory when Allis-Chalmers took over in 1931.

Below: The 6A couldn't save Advance-Rumely from takeover.

Rushton

UK

Specifications for Rushton General
Engine: Water-cooled, four-cylinder
Bore x stroke: 4.125 x 5.0in (100 x 125mm)
Capacity: 267ci (4,165cc)
Power: Not known
Transmission: Three-speed
Speeds: 1.3-6.8mph (2.1-10.9km/h)
Weight: c.2,800lb (1,260kg)

George Rushton was an Englishman who sincerely wanted to design and produce an agricultural tractor. Without the wherewithall to do this himself, he approached AEC, the long-established bus and truck manufacturer in London. With one eye on the expanding tractor market (this was 1926, and the Wall Street Crash and resulting slump was still a few years away) AEC agreed, and took Mr Rushton on.

The experimental Tri Tractor was the result, but a more accurate pointer to the future came when the company bought a Fordson for evaluation. The Fordson-like General tractor was announced in July 1928, using some existing AEC components. It was slightly heavier than the Fordson, a little more powerful and a lot more expensive.

Rushton planned to buy components direct from AEC, and 500 sets of parts were ordered, to get production going. But by this time, AEC had had a change of heart and wanted to pull out. By floating shares, George Rushton found the money to pay off the tractor's development costs and start production at AEC's Walthamstow factory. Despite this difficult birth, the Rushton tractor did get into series production, and the 14/20 pictured here clearly shows its Fordson inspiration. A crawler version was later produced alongside it. But neither lasted long, and the company went into receivership in 1932.

Below: The Fordson-like Rushton was short-lived.

SAME Silver 90

ITALY

Specifications for Silver 90 (1995)
Engine: Water-cooled, four-cylinder, turbo
Capacity: 4.0 litres
Power: 90hp @ 2,500rpm
Transmission: Agroshift

Question: Which country came up with the first four-wheel-drive tractor, and which the first diesel? Most of us would probably answer the USA and Germany, respectively, but we'd be wrong on both counts.

Francesco Cassani, of Bergamo province in Italy, applied the diesel engine to a tractor in 1924. And he built (and sold) the world's first four-wheel-drive tractor in 1952. By then, Snr Cassani had set up his own company – Societa Anonima Motori Endotermici – SAME. Several innovative machines followed, one of the earliest being a little 10hp device

with a reversible seat, so that the driver could keep a check on rear-mounted implements.

After taking over Lamborghini tractors in 1972, and the Swiss concern Hurlimann in '77, SAME later merged with Deutz-Fahr, and now produces a range of up to the minute machines. Some are mini-tractors, fitted with either Mitsubishi diesels or the firm's own air-cooled three-cylinder motor. But for the full-size machines, SAME announced its new 1000 series diesel in the mid- 1990s, based on the modular principle in which a whole range of power units share many common components. Unusually for a modern unit, it is air-cooled. The Silver pictured here covered the 80-100hp class in the late '90s, with 256ci (4.0 litres) four-cylinder or 385ci (6.0 litres) six-cylinder versions of the SAME 1000 engine.

Below: SAME was a four-wheel-drive pioneer.

Samson

USA

Specifications for Model M (1920)
Engine: Water-cooled, four-cylinder
Bore x stroke: 4in x 5.5in (100 x 138mm)
Capacity: 276ci (4,306cc)
PTO power: 19hp
Drawbar power: 11.5hp
Transmission: Two-speed
Speeds: 2.3 and 3.2mph (3.7 and 5.1km/h)
Fuel consumption: 6.9hp/hr per gallon
Weight: 3,300lb (1,485kg)

In the North American car market, Ford and General Motors were bitter rivals. So when it became clear that Ford was serious about building a tractor, General Motors had no choice but to follow suit. Ford's tractor would be mass-produced, just like his Model T; it would be cheap, light and reliable.

But there was a difference. Henry Ford came from farming stock – for him, it was partly a labour of love. He knew what small farmers needed from a tractor, and had been pondering the problem for several years. Many prototypes had been built before the classic Fordson was finalised, and announced to the world in 1917. For General Motors, tractors were no more than a business opportunity, one that they were following only because they'd get left behind if they didn't.

GM didn't have time for years of R & D and needed a ready-made tractor. It appeared as the Samson, an established manufacturer from Janesville, and GM bought them up in February 1917. The new Samson M was a conventional four-wheel, four-cylinder tractor to meet the Fordson head-on, but it didn't last long. In the difficult years after World War I, GM was forced to shut down its unprofitable sections – Samson succumbed in 1922

Below: Where Henry Ford went, General Motors had to follow. But GM lacked Ford's in-depth tractor R&D, not to mention his farming experience.

Saunderson Model G

UK

Specifications for Saunderson L
Engine: Air-cooled, single-cylinder
Bore x stroke: 6.5 x 10.0in (163 x 250mm)
Capacity: 332ci (5,174cc)
Transmission: Two-speed

Mr. H. P. Saunderson of Bedford, England, had travelled to Canada early in life, returning home in 1890 with a keen interest in mechanised farming. At first, he imported Massey-Harris machinery, but began producing vehicles himself from around 1903. Three and four-wheeled models were available, such as the four-wheel, single-cylinder Model L. These were dual-purpose machines, with their load beds over the rear wheels aimed as much at load carrying as field work.

Sales were low until Saunderson began making pure farm tractors, the Universal series in general and the Model G – as pictured here – in particular. The G was a more conventional machine than earlier Saundersons (some of which were designed to fit within the shafts of horsedrawn machinery. The arrival of the G in 1916 coincided with Britain's wartime boom in tractor demand, which also established the Fordson. For a short time, the Saunderson was Britain's best selling tractor, and unlike many of its pre-war rivals survived after 1918, alongside newer tractors like the Austin. A French version was built under the Scemia name. Sadly, Saunderson did not survive the tractor slump of the 1920s.

Below: Saunderson enjoyed Britain's brief tractor boom of 1916-18.

Sawyer-Massey

???

Specifications for Sawyer-Massey 20-40
Engine: Water-cooled, four-cylinder
Bore x stroke: 5.625 x 7.0in (141 x 175mm)
Capacity: 692ci (10,795cc)
PTO power: 40hp
Drawbar power: 20hp (nominal)
Transmission: Two-speed
Speeds: 2 & 3mph (3.2 & 4.8km/h)
Weight: 11,800lb (5,310kg)

Despite its name, Sawyer-Massey was never part of Massey-Harris, or Massey Ferguson for that matter, but they were related. The Sawyer company was established in 1836, and built a range of steam traction engines, and later kerosene tractors and stationary engines as well. In 1892, the Massey family bought 40 per cent of its shares, and maintained a financial interest until 1910 – during this time, the firm became known as Sawyer-Massey.

Sawyer-Massey tractors were used in both agricultural and road making, and in the usual way of the period, ran on kerosene but were started on gasoline. Information is scarce, but we do know that their smallest model was a 10-20, using a four-cylinder Waukesha engine of 326ci (5,086cc). It also weighed 5,400lb (2,430kg), so it's clear that the company had little interest in the new breed of lightweight tractors! A mixture of S-M's own engines and bought-in units powered Sawyer-Massey tractors – as well as the Waukesha, both Erd and Minneapolis motors were used. All were big, heavy machines that owed as much to steam traction engines as to more modern thinking. The 25-50 for example, was a 6-8-plough machine that weighed 17,500lb (7,875kg), and used a massive, slow-turning four-cylinder engine made in-house by Sawyer-Massey. There was also massive 30-60 to rival the Advance-Rumeley Oil Pull and others (not to mention steam engines) but not many of these were made.

Below: The big Sawyer-Masseys owed much to steam traction engines.

SFV/Vierzon 551

FRANCE

No specifications available

The French tractor industry has come up with fewer globally-known makes than most. Everyone has heard of Renault, but not so many would list Citroen as a tractor maker – the Citroen-Kegresse half-track was based on a Citroen car. Likewise, the Latil four-cylinder tractor, an early four-wheel-drive machine, was on sale in 1930; in the '20s, there was a French made version of the Austin tractor; and after World War II, the Simca-produced Someca appeared.

But there's another French manufacturer. The Societe Francaise de Materiel Agricole et Industriel Vierzon (known more conveniently as SFV) was formed in 1935. Throughout its 25-year history, SFV concentrated on single-cylinder semi-diesels. This was a European form of machine: Lanz of Germany, Landini in Italy and Marshall of England all produced similar machines from the 1920s to late 1950s. Instead of a multi-cylinder gasoline or diesel engine, the semi-diesel used just one big cylinder. Compared to a four-cylinder of similar power they were crude, rough and vibratory, but had the saving grace of simplicity and complete reliability. The SFV 551 shown here was typical.

Right: A little crude, but simple, cheap and reliable – the SFV.

D
SFV

SFV 302

FRANCE

Despite its green and yellow livery, this SFV had no connection whatsoever with John Deere. But it's marked out as an SFV by the patriotic brass and enamel badge combining a castle and the French tricoleur. Like other SFVs, the 302 shown here used a horizontal, single-cylinder semi-diesel. Not a true diesel then, but one which relied on a hot spot on the cylinder-head, rather than compression, to ignite the fuel/air mixture. It also operated on two-stroke principles, with no valves. Starting was an art in itself, using either a cartridge system or a blowlamp to heat the hot spot. The SFV's cylinder-head was easily accessible, poking out of the front of the machine.

The result was a somewhat crude and smokey engine, but one which was supremely simple and reliable. They would also run happily on very low grade fuel. This philosophy worked fine until the late 1950s, when farmers were demanding more sophistication from their tractors. By 1960, all the semi-diesels had gone. In England, Marshall stopped making agricultural tractors altogether, while Lanz of Germany was taken over by John Deere. As for SFV, it was bought up by Case, while the factory at Vierzon was later bought by Ford.

Below: Unlike Landini, SFV failed to leap into the multi-cylinder era.

Silver King

USA

Specifications for 04 (1936)
Engine: Water-cooled, four-cylinder
Bore x stroke: 3.25in x 4.0in (81 x 100mm)
Capacity: 133ci (2,075cc)
PTO power: 19.7hp @ 1,400rpm
Drawbar power: 10.5hp
Transmission: Four-speed
Speeds: 2.3-14.5mph (up to 25mph) (3.7-23.2km/h, up to40km/h)
Fuel consumption: 8.4hp/hr per gallon

The evocatively named Silver King tractors were built by a company with a far less romantic name – the Fate-Root-Heath Company of Plymouth, Ohio! They were small machines, and the three-wheel tested by the University of Nebraska was typical. The bought-in Hercules engine produced just under 20hp at

1,400rpm. It was a modest output, but the high fourth gear allowed speeds of up to 25mph. This, needless to say, was with the optional pneumatic tyres, not on steel wheels! Nebraska tested this Silver King on both steel and rubber, which gave fuel economy figures of 4.37 and 7.31hp/hr per gallon respectively, underlining the far greater efficiency of pneumatics. Unfortunately, this tractor also suffered from overheating during its Nebraska test, thanks to the unshrouded radiator painted in aluminium. The addition of a shroud and an unpainted radiator solved the problem.

After World War II, Silver King offered another three-wheel row-crop tractor, but this time with a slightly bigger Continental engine of 162ci (2,527cc).

Below: Like Minneapolis-Moline, Silver King provided a high top gear ratio for faster on-road speeds.

Someca 880

FRANCE

Specifications for DA50 (1953)
Engine: Water-cooled, four-cylinder
Bore x stroke: 3.94in x 4.72in (99 x 118mm)
Capacity: 230ci (3,588cc)
PTO power: 38.9hp @ 1,500rpm
Torque: 210lb ft (165Nm) @ 992rpm
Drawbar power: 26.6hp
Transmission: Five-speed
Speeds: 2.3-13.3mph (3.7-21.3km/h)
Fuel consumption: 13.34hp/hr per gallon
Weight: 5,011lb (2,275kg)

Ask any tractor enthusiast to name three famous French tractor makers, and many will have difficulty getting beyond Renault. But there were others, the only surprise being that France, with its large agricultural sector, did not develop

a larger and more significant tractor industry than it did. In the early years, it was less vulnerable to the Fordson revolution, thanks to tariff barriers – that was precisely why the English Austin tractor was built in France for the French market, to avoid import duties.

The Someca was an attempt to produce a homegrown tractor after World War II, a co-operation between the Simca car company, and an established tractor manufacturer. Some were exported to the USA, and in fact a Someca was the first French tractor ever tested at Nebraska. Not the 880 pictured here, but the DA50 of 1953, which used a four-cylinder diesel engine from OM Milano of Italy. It produced 39hp at the PTO, and drove through a five-speed gearbox, with 210lb ft (155Nm) at just under 1,000rpm. The later SOM-45 used a 253ci (3,947cc) Milano diesel of 36hp.

Below: Despite its many small farms, France did not develop a major tractor industry. The Someca did not survive.

Steiger

USA

Douglas and Maurice Steiger farmed 4,000 acres near Red Lake Falls, Minnesota, but couldn't find the tractor they wanted – so they built their own. Using readily available components, they assembled Steiger 'No1' in the winter of 1957/58. With four equal-sized wheels, all of them driven, an articulated drive shaft and a big, powerful diesel engine, all the classic ingredients of the massive Steiger tractor were there from the start.

The Steigers went on to build 120 big tractors between 1963 and '69, assembling them in their dairy barn. In 1969, they joined forces with a business

consortium, and moved production to a disused tank factory in Fargo, North Dakota. As well as selling direct to customers, Steiger did good business in making giant 4x4 machines for the mainstream manufacturers. They long had links with Ford, and the Allis-Chalmers 440 (see next page) was simply a rebadged Steiger. The company was sold to Case in 1986. Since then, all Steigers have been badged Case-Internationals.

Below: The Steiger brothers invented a new type of tractor.

Steiger Bearcat/AC440

USA

Engine: Water-cooled, V8
Capacity: 555ci (8.7 litres)
Transmission: 10-speed

The AC440 was nor more nor less than a Steiger Bearcat in Allis-Chalmers colours. For A-C, this was a relatively quick and simple way into the big tractor market, after its attempts at an in-house 4x4 ended in failure. The Bearcat was powered by a Cummins V8 diesel of 555ci (8.7 litres) and used a 10-speed transmission. About 1,000 were sold as AC440s between 1972 and '76, when Allis launched its own four-wheel-drive machine.

By this time, Steiger was well established as the leading American maker of big tractors. They hadn't had an obvious plan for doing so, but neighbours liked the look of their first machine, and asked for copies. As it happened, the first customer-request Steigers were a little smaller, more conventional and civilised than that first one. 'It was a a power beast, but a numb lump,' said tractor historian Peter Simpson. 'It's tiller steering made it difficult to drive and Steiger's prospective customers would want something more manageable.'

So the Steiger 'No2's, which evolved into the 1200, were powered by a smaller 3.71 Series Detroit Diesel of 118hp. They had a conventional steering wheel and were easier to operate, but really a new class of tractor had been born.

Right: Badged an Allis-Chalmers, but really a Steiger Bearcat.

Steiger ST310

USA

Specifications for Panther III ST-325 (1977)
Engine: Water-cooled, six-cylinder, turbo-intercooled
Bore x stroke: 5.4in x 6.5in (135 x 163mm)
Capacity: 893ci (13,931cc)
Power: 270hp @ 2,100rpm
Transmission: Ten-speed
Speeds: 2.3-19.9mph (3.7-31.8km/h)
Fuel consumption: 14.82hp/hr per gallon
Weight: 31,080lb (13,986kg)

The Steiger brothers loved to name their machines after big cats – the associations with power and strength were obvious. So the ST310 shown here carried the model name 'Panther.' The Bearcat was later built for Allis-Chalmers as the more prosaic 'AC440' and there were Pumas, Cougars, Lions and Tigers as well. But it was more than just marketing – with massive, torquey diesel engines of up to 525hp, Steigers could claim to be the most powerful tractors of their time. And they were well thought of. In at least one survey of farmers, the Steiger name was cited as the most popular and well-known brand of 4x4 tractor on the market.

The Panther ST310, pictured here in the traditional Steiger lime green, used a six-cylinder Cummins engine of 855ci (13,338cc), later upgraded as the ST325 with an 893ci (13,931cc) Cummins delivering 270hp at 2,100rpm. To give an idea of the scale of these tractors, the 325 weighed in at 31,080lb (13,986kg) – over 15 tons!

Right: Steiger used Detroit Diesels and Cummins engines.

AC 440

BISSEN BROS.
PANTHER
ST310
STEIGER

Steiger 1360

USA

Specifications for Panther CP-1360 Diesel (1982)
Engine: Water-cooled, six-cylinder, turbo-intercooled
Bore x stroke: 5.5in x 6.0in (138 x 150mm)
Capacity: 855ci (13,338cc)
PTO power: 326hp @ 2,100rpm
Drawbar power: 289hp
Transmission: Twelve-speed
Speeds: 2.4-16.5mph (3.8-26.4km/h)
Fuel consumption: 15.45hp/hr per gallon
Weight: 35,480lb (15,966kg)

The Panthers of the 1970s weren't the biggest and most powerful Steigers ever built. Back in 1963, the Steigers had started work on their second generation of tractors. Promoted as 'The Big Ones', they all used off-the-shelf parts: Detroit Diesel engines, Allison or Spicer transmissions, Clark axles with planetary final drives and of course Steiger's own swinging power divider, which articulated the drive shaft and was patented. The smallest of the second

generation was the 1250, with a 127hp Detroit four-cylinder engine. Next up, the 1700 and 2200 both used V6 Detroits, driving through nine-speed gearboxes – power figures were 216hp and 265hp, respectively. The last of the second generation to appear was also the biggest, the 3300. This time, the Detroit Diesel was a V8, which generated 350hp and used a 16 x 8 transmission. Up and running, the 3300 could work 38 acres an hour with a 56ft chisel plough – quite something in 1967.

By 1969, Steiger was about to lose its status as a family owned firm, and the Tiger 800-series of that year turned out to be the last tractor dseigned and built on the farm. For the first time, the brothers had forsaken Detroit Diesel power for Cummins, an engine supplier (along with Caterpillar) they would use for the next 30 years, though some Steigers would use Caterpillar diesels. The 1360 shown here was typical of an early '80s Steiger, produced in the last few years of the company's independence.

Below: Later Steigers were powered by Cummins diesels.

Steiger/Case International 9250

USA

Specifications for Case International 9350 (1996)
Engine: Water-cooled, six-cylinder turbo-diesel
Capacity: 532ci (8.3 litres)
Power: 310hp
Transmission: Twelve-speed Powershift

When Case IH took over Steiger in 1986, it may well have intended to rebadge all Steigers as Case models, and cut costs by bringing production into one of Case's own factories. But to do that would have wasted a famous name, as Steiger was viewed as the leading four-wheel-drive tractor brand in North America.

So the 9250 pictured here was officially a Case International 9250, complete with the red and black Case colour scheme. But after letting it drop for a few years, Case revived the Steiger name for the new 9300 series of the 1990s. Steiger tractors are now painted in Case IH colours, but they're still made in Fargo, North Dakota, and still carry the Steiger badge. The 9200 series had continued the Steiger tradition of a massive four-wheel-drive tractor, using an articulated chassis, and now with the option of triple wheels – that meant twelve tyres to buy at replacement time!

The 9300 continued the same layout, but by the late 1990s offered a full range of ten machines, ranging from the 240hp 9330 to the 425hp 9390,

Steiger 9380

USA

SPECIFICATIONS
Steiger 9380 Quadtrac (2002)
Engine: Water-cooled, six-cylinder, turbo-intercooled
Bore x stroke: 5.6 x 6.1in (140 x 152mm)
Capacity: 855ci (14.0 litres)
Power: 360hp
Transmission: Twelve-speed SyncroShift
Track width: 30in (750mm)
Weight: 43,750lb (19,688kg)

'By properly selecting tyres and adjusting tyre pressures,' went Case's publicity for the 9300 series, 'the tractor's weight can literally float across the soil for reduced compaction.' Well, perhaps. No-one disputed that modern tyre technology, and double- or triple-wheel options, had increased traction while reducing compaction. But for the ultimate in spreading a tractor's weight, you still couldn't beat tracks, and Case recognised this with the 9380 Quadtrac shown here.

Caterpillar had pioneered modern rubber tracks, to give a better compromise between hard-top speed and quagmire traction. As its name suggested, the Steiger Quadtrac used four of them, one in place of each wheel. This wasn't a new idea – the Roadless Traction company had offered a four-track conversion for the four-wheel-drive Land Rover back in the late 1950s.

But the Quadtrac was a very different beast to that one. Each one of its rubber tracks was 30 inches wide, giving 68.5 inches of contact with the ground. The tracks could pivot up and down by ten degrees, allowing them to

Above: Case International 9250, but the Steiger name would revive.

including two row-crop models. Engines were a choice of Case's own 532ci (8.3-litres) (and the Cummins N14 or M11. A 12-speed transmission was standard, with 24-speed SynchroShift and 12-speed Powershift options. Lift capacities ranged from 14,689lb-19,586lb (6,662-8,884kg).

Above: Best of both worlds?

follow ground contours more easily, and each of the four worked independently, having a better chance of keeping in contact with the ground. Twenty-six degrees of oscillation between the front and rear of the Quadtrac helped as well, while improving comfort for the driver.

Despite all of this, Case claimed that the Quadtrac would steer just like a conventional four-wheel-drive tractor, with a turning radius of less than 20 feet. And compared to a two-track unit, it was able to keep the tractor's articulation, so you could still turn under load. Power came from a 360hp Cummins N14, with turbo and after-cooling, while transmission options were the same as on the rest of the 9300 series.

Terra-Gator 8103

USA

Engine: Water-cooled, six-cylinder
Capacity: 519ci (8.1 litres)
Power: 320bhp
Transmission: 11-speed powershift

Terra-Gators are specialist high capacity spreaders, whose massive low pressure tyres are designed to impose as little ground pressure as possible, to minimise earth compaction. The 1664 pictured here is an earlier example, but in September 1998 the company announced the new 8103.

This used a six-cylinder John Deere diesel. Turbocharged, and of 519ci (8.1 litres), this produced 320hp. Transmission was taken care of by an Ag-Chem Terra-Shift unit, allowing for clutch-free changes on the move and providing a range of

close ratios. The gear ratios were so close that the steps between them were as little as 14.5 per cent. A new cab on the 8103 improved visibility, and there were better ergonomics and less noise. Reportedly to increase the unit's life span and improve reliability, a John Deere Team Mate 2 1400 series axle was added. The chassis was new as well, a rectangular tube frame for greater strength and flexibility, and to reduce the effects of stress.

The 8103 was just one of a range of Terra-Gators, and by no means the biggest – the 1903 used a 400bhp Caterpillar diesel – and was joined by a four-wheel-drive version, the 8144, in 2000. Ten thousand Terra-Gators of various types have been built.

Below: Tractor or truck? The Terra-Gator is a specialist spreader.

Turner V4

UK

Engine: Water-cooled, V-4
Bore x stroke: 3.75 x 4.5in (94 x 113mm)
Capacity: 207ci (3,227cc)
Power: 34hp
Transmission: Four-speed

'For the greatest pull on earth...on wheels...at the lowest fuel cost. The Turner "Yeoman of England" diesel tractor gives you not only great work capacity but more working days.' So went the publicity for Turner's new diesel tractor of 1949 – Turner claimed it would give the pull of a crawler with the economy of a wheeled tractor. Sadly, it was oversold.

Turner, based in Wolverhampton in the English Midlands, was a respected, long established maker of winches and stationary engines. Looking around for

new product lines after World War II, it decided to branch into diesel engines. A family of three was offered, a single-cylinder, V-twin and V-4, all with the same 3.75in bore and 4.5in stroke. It was the 4V95 V-4, with its 68 degree cylinder angle, that formed the basis of the Turner tractor. With 34hp, the V-4 seemed just the right sort of size for a small machine.

The prototypes were extensively tested on farms, but in service, the V-4 proved woefully unreliable. By its looks and exhaust note, the Turner V-4 seemed powerful – unfortunately, it wasn't that either. Turner persevered with its tractor for eight years, but sold fewer than 2,500 in that time. It gave up on tractors, concentrating instead on making transmissions.

Below: Sought-after by collectors today, the Turner V4 failed to live up to the promise of its engine layout or exhaust note.

Universal 530

ROMANIA

Specifications for Long 510 DTC (Universal)
Engine: Three-cylinder, water-cooled turbo-diesel
Bore x stroke: 4.0 x 4.3in (100 x 108mm)
Capacity: 165ci (2,574cc)
PTO power: 49.4hp @ 2,400rpm
Drawbar power: 40hp
Transmission: Eight-speed
Speeds: 1.4-14.3mph (2.2-22.3km/h)
Weight: 5,625lb (2,531kg)

Looking back over the history of tractors, innumerable manufacturers have registered under the name 'Universal'. For hard nosed farmers, it's a good name, suggesting a tractor that can turn its hand to any job that needs doing.

There have been Universals from both the USA and Britain, but this one hailed from Brasov in Romania. In North America, it was sold by the Long Manufacturing Company of North Carolina, which started out producing its own machines in the late 1940s, but turned to importing the Romanian tractors in order to stay competitive. The 530 illustrated here was not tested at Nebraska, but the similar 510 was, in both naturally aspirated and turbo-diesel forms. Both used Universal's own three-cylinder diesel, measuring 165ci (2,574cc), but oddly, Nebraska recorded near-identical PTO power figures for turbo and non-turbo! Front-wheel-assist was optional on the turbo. There was also a 610, using a four-cylinder 219ci (3,416cc) version of that same engine, giving just over 64hp at the PTO. All three were offered in 1981.

Below: The Romanian Universal was one of several Eastern Bloc tractors.

Ursus C-325

POLAND

Specifications for C-325 (1961)
Engine: Water-cooled, twin-cylinder diesel
Bore x stroke: 3.9 x 4.7in (98 x 118mm)
Capacity: 111ci (1,732cc)
PTO power: 24.6hp @ 2,000rpm
Drawbar power: 18hp
Transmission: Six-speed
Fuel economy: 14.78hp/hr per gall
Weight: 4,849lb (2,182kg) (inc test ballast)

Ursus of Poland has a long history, but not always with tractors. It was set up in Warsaw in 1893 to produce fittings for the sugar and food industries. By 1913, it produced only internal combustion engines – 6,000 were produced before the outbreak of World War I, some of up to 450hp, and in 1918 a

prototype tractor engine was built. The first complete machine was built in 1922, and just one hundred of these were built over the next five years – by 1929,the company was also making buses and trucks.

Nationalised after a commercial collapse in 1930, the company thrived during World War II, building cars, motorcycles, buses and tanks, plus about 700 military tractors. It also carried on producing power units, for planes, agriculture and stand-alone units. After the war, production restarted with a new tractor based closely on the German Lanz Bulldog – 60,000 of these were built between 1947 and 1959. Meanwhile, the company developed its own machine, the twin-cylinder diesel C-325 – it was the first of a long line of light tractors that could be traced up to 1993.

Below: In the 1960s, Ursus developed its own range of small tractors, but until 1959 produced a machine based on the single-cylinder Lanz.

Ursus 4514

POLAND

Specifications for C-350 (1968)
Engine: Water-cooled, four-cylinder diesel
Bore x stroke: 3.7 x 4.3in (93 x 108mm)
Capacity: 185ci (2,884cc)
PTO power: 43hp
Drawbar power: 37hp
Transmission: Ten-speed
Speeds: 0.9-15.9mph (1.4-25.4km/h)
Fuel consumption: 14.6hp/hr per gall
Weight: 4,880lb (2,196kg)

During the '60s, Ursus became increasingly dependent on Zetor, with about half its components coming direct from Czechoslovakia. It was actually a two-way trip, since the Zetor's front axle and hydraulics were built by Ursus. But there could be no doubt that Zetor was far ahead of Ursus technically.

One writer, in the *Vintage Tractor Album*, commented after the Ursus agreement with Massey-Ferguson: 'One hopes that this influx of Western echnology improves the Poles' metallurgical skills, because the quality of the netal in both Ursus and Zetor gears has sometimes given ample grounds for criticism and caused Zetor men in unguarded moments to mutter that they wish they could return to producing the whole tractor themselves.\

In fact, the 'new' Ursus 4011 of 1965 was, according to one writer, the already-obsolete Zetor 4011. The 4514 pictured here was a development of that machine, though it went to be updated, and as the C-355 and C-360 was actually produced up to 1992.

But things were already changing in the early 1970s, when a licencing agreement with Massey-Ferguson led to a new range of 38-72hp machines. Ursus was privatised in 1998, and now faces a tough post-communist world.

Below: Don't be fooled by the badge: it's an Ursus, albeit Zetor-influenced.

Valmet 6300

Specifications for Valtra S Series (2002)
Engine: Water-cooled, six-cylinder turbo-diesel
Capacity: 538ci (8.4 litres)
Power: 26hp @ 2,200rpm
Torque: 1,050Nm @ 1,400rpm
Tranmission: 40-speed F and R
Speeds: Up to 31mph (50km/h)
Weight: 19,360lb (8,800kg)

Unlike many of the manufacturers featured here, Valmet of Finland didn't get into the tractor business until after World War II. In 1952, Valtion Metallitehtaat (soon shortened to Valmet) began production in a former rifle factory, whose workers had the necessary precision engineering skills.

The little Valmet 15, a 15hp 1.5 litre side-valve machine, found a ready market in postwar Finland, and was soon followed by a more powerful version. Valmet's first diesel, the 33D, appeared in 1957, powered by the company's own three-cylinder engine. It used direct injection, and was water-cooled. There were those who wanted air-cooling to avoiding freezing in the harsh Scandanavian winter. But Finland has hot summers as well, so the 33D was water-cooled. Four-cylinder diesels followed, and the new 900 of 1967 paid fresh attention to driver ergonomics, until then a somewhat neglected aspect of tractor design. It also had an integral safety cab with rubber mounting.

Six-cylinder and turbocharged diesels took Valmet into higher-power markets, and in fact there's a difference of opinion as to whether the Finns or Swedish Volvo built the first European turbo-diesel tractor.

Right: From humble beginnings, Valmet has become a significant European manufacturer.

Valmet
6300

Versatile 895

CANADA

Engine: Water-cooled, six-cylinder, turbo-intercooled
Bore x stroke: 5.5 x 6.0in (138 x 150mm)
Capacity: 855ci (13,338cc)
Drawbar power: 252hp @ 2,100rpm
Transmission: Twelve-speed
Speeds: 2.6-14.3mph (4.2-22.9km/h)
Fuel economy: 15.15hp hr per gall
Weight: 24,620lb (11,079kg)

Just as the Steiger brothers were developing their super-tractors in Minnesota, so a Canadian company was doing the same in Winnipeg. The essential ingredients were the same as well. Four-wheel-drive through four equal-size wheels, articulated steering and a big diesel engine with more power and torque than any conventional tractor. And the two companies that built

them would become known all over the world.

The Hydraulic Engineering Company started out in Toronto in 1947, moving to Winnipeg three years later. It made small agricultural implements at first, moving up to a big self-propelled swather in 1954. In 1963, it went public and took on a new name – Versatile.

Versatile's first super-tractor was quite small by Steiger standards. Unveiled in 1966, the D100 was powered by a 125hp British Ford industrial engine, a six-cylinder diesel. Only one hundred of these were made before it was replaced by the D145.

Like Steiger, Versatile tractors were designed for the big open spaces of the American Mid-West, and Canada. And, also like Steiger, Versatile became known as one of the leading makes of four-wheel-drive super-tractor. The 895 pictured here was a typical such machine of the late 1970s.

Below: Versatile was the Canadian Steiger.

Versatile 946

CANADA/USA

Engine: Water-cooled, six-cylinder, turbo-diesel
Capacity: 855ci (13,338cc)
Power: 325hp @ 2,100rpm
Transmission: 24-speed
Weight: 14,500kg (31900lb)

In 1986, Versatile closed down, due to financial problems with its parent company, Coronet Industries. But this certainly wasn't the end of the Versatile story. As the Fiat marketing deal had shown, the super-tractor market was one that mainstream manufacturers wanted a slice of. And, as ever, it was much quicker and cheaper to bolt your badge onto an established product, than develop your own. That's why Case IH took over Steiger in 1986, and why Ford New Holland did the same for Versatile the following year. Within a few

months, tractors were rolling out of the Winnipeg plant once again.

Now in Ford NH colours (the Versatile name was retained in small print) the 6-series went on sale in 1987, and was offered for six years. The 946 pictured here was the second largest machine in the range – the 846 and 876 were lower powered,the 976 higher – it was also the only 6-series machine to be exported to Britain. Like most Versatiles, it was powered by a Cummins six-cylinder diesel, the 855ci (13,338cc) in this case, turbocharged and aftercooled to produced 325hp. At least one British farmer found it to be very economical as well – Rex Sly's 946 consumed just over a gallon of fuel per acre. With a 239-gallon tank, that could mean four days, work without thinking about filling up.

Below: Versatile was saved by Ford, and its name lived on (in small print).

Versatile 256

CANADA/USA

Engine: Water-cooled, four-cylinder, turbo
Bore x stroke: 4.0 x 4.7in (100 x 118mm)
Capacity: 239ci (3,728cc)
PTO power: 84hp @ 2,500rpm
Transmission: Auto hydrostatic + three-speed range
Speeds: 0-19.9mph (0-31.8km/h)
Fuel consumption: 16.9hp hr per gall
Weight: 9,150lb (4,118kg)

Most super-tractors are in the 200-400hp range, though in 1977 Versatile did unveil a prototype eight-wheeler which broke all size records. The 1080 'Big Roy' had all eight of its wheels driven, and was powered by a 600hp Cummins diesel. It never went into production, and is now preserved in the Manitoba Agricultural Museum.

The 256 pictured here did go into production in the early '80s, but by

Versatile standards it was a baby. Only four cylinders and 239ci (3,728cc) – though it was turbocharged – seemed like some sort of mistake. It even weighed less than 10,000lb! In appearance though, the 256 was styled like a miniature super-tractor, and it did have four-wheel-drive through equal-sized wheels. More to the point, it had hydrostatic transmission, which was fully automatic, giving infinitely variable ratios from zero to 19.9mph. there was a three-speed manual transmission as well, to give three ranges.

At the same time, the bigger Versatiles were being sold in some markets with Fiat badges and colours. A marketing agreement with the Italian giant in 1979 allowed for a four-model range to be sold outside the US, Canada and Australia – the Fiat 44-28 was a 280hp tractor, powered by a Cummins 'Constant Power' with 31 per cent torque back-up.

Below: Some Versatiles were later badged as Fiats.

Versatile/New Holland TV140

CANADA/USA

Specifications for New Holland TV140
Engine: Water-cooled, six-cylinder turbo
Capacity: 456ci (7.5 litres)
Power: 105hp
Transmission: Auto hydrostatic + three-speed range
Speeds: 0-18mph (0-28.8km/h)

Ford New Holland merged with Fiat in 1991, after which Versatiles were badged as New Hollands only. Eight years later there was another merger, this time with Case. As part of the deal, the US Dept of Justice ruled that the Winnipeg plant be sold. It was bought by Buhler Industries in 2000.

So the TV140 shown here isn't produced at Winnipeg. It was launched in 1998 to replace the ageing 9030 Bi-directional. Like that tractor, the TV140 can

be operated in either direction, with the operator facing forwards, out of the cab or over the engine – hence the name. The Turnabout console swivels by 180 degrees, taking all the controls and instrumentation with it, doing all this in less than five seconds. And just like the earliest Versatiles, there is articulated steering, which in this case gives the TV140 a best-in-class turning with a 70-inch (1,750mm) wheel track.

Badged now as a New Holland, the early TV140 was powered by a 135hp six-cylinder diesel, with 42 per cent torque rise. A municipal Utility version was introduced in early 1999, with a 105hp version of the same engine. Styling was deliberately skewed to fit in with the equivalent New Holland conventional tractors – the same sloping bonnet as the Genesis 70 and Gemini 60 for good visibility.

Below: Versatile's name is just visible on this latest New-Holland.

Volvo BM 650 Turbo

SWEDEN

Specifications for Valtra Mega 6200 (latest equivalent of Volvo BM650 Turbo)
Engine: Water-cooled, four-cylinder, turbo
Power: 80hp @ 2,225rpm
Torque: 236lb ft @ 1,400rpm
Transmission: 12-speed (inc shuttle, 36-speed option)
Speeds: 0.4-25mph (0.6-40km/h)
Weight: 8,932lb (4,060kg)

Swedish manufacturer Volvo is well known for its range of buses, trucks and cars, but has also been making tractors since the 1920s. Now part of the Valmet (recently renamed Valtra) group, Volvo took over Bolinder-Munktell, and was later taken over itself by Finnish tractor maker Valmet. The latter began as a co-operative deal, when the two firms worked together to build the Nordic range of tractors, specifically for Scandanavian conditions.

Volvo first came to diesel power in the 1950s, using a Perkins L4 engine in

its T30 tractor. It wasn't long of course, before the Swedes began making their own diesels. Typical was the 95bhp four-cylinder turbo-diesel of the 1980s, which powered the two-wheel-drive 805 and four-wheel-drive 805-4. There was also a smaller 405 and the much larger 2105.

In keeping with its safety-first image, Volvo was one of the first tractor manufacturers to design and fit safety cabs. Tractors flipping over backwards or rolling sideways often had fatal results, but it wasn't until the early 1950s that manufacturers started examining ways of making tractors safer. Before the advent of the safety cab, there were various attempts to do this, such as switches which automatically cut the engine if the driver left his seat or the tractor tilted more than certain extent. Astonishingly, one early 'safety' frame, demonstrated at Nebraska in 1952, protected the tractor in a roll-over, but left the driver exposed! Not until the late 1960s did effective roll-over protection (ROPS) become common.

Below: Car, bus and truck maker Volvo has long built tractors as well.

White

USA

It's familar now as part of the AGCO empire, but White's tractor heritage is relatively short. White was an American maker of big rig trucks, but was determined to get into the tractor market. It did so by buying Oliver in 1960, Cockshutt in '62 and Minneapolis-Moline the following year. In 1969, the ranges were merged, sharing many components, though the traditional badges and colour schemes were kept on to make the most of residual loyalty to these long-established names.
'In 1974,' wrote auther P. W. Ertel, 'White dropped the pretence that its Oliver,

Cockshutt and Minneapolis-Moline lines were different tractors, and consolidated them all under the White name.' And, he added acerbically, 'White proved more adept at finishing off tractor companies than growing them.' White of course, soon faced its own problems, and after a couple of takeovers, the White name is being used in just the same way – poetic justice?

Below: Truck maker White bought its way into the tractor business.

White 4-150

USA

Specifications for 4-180 (1975)
Engine: Water-cooled V8 diesel
Bore x stroke: 4.5 x 5.0in (113 x 125mm)
Capacity: 636ci (9,922cc)
PTO power: 181hp @ 2,800rpm
Drawbar power 157hp
Transmission: 12-speed
Speeds: 1.9-18.9mph (3.0-30.2km/h)
Fuel consumption: 11.56 hp hr/gal
Weight: 20,600lb (9,270kg)

Nineteen seventy-four was an important year for White. That was when the old respected badges of Oliver, Minneapolis-Moline and Cockshutt were finally dropped – from now on, all tractors would bear the White name.

It certainly came in with a bang. White's first own-brand tractor was the big four-wheel-drive 4-150 Field Boss – '4' related to the number of driven wheels, '150' to the approximate PTO horsepower. Actually, the 4-150 wasn't as new as it seemed. Aiming for the articulated tractor market, it consisted of two Oliver power trains, hinged in the middle with power steering. Power came from a 636ci (9,922cc) Caterpillar V8 diesel.

What made the 4-150 unusual was its low profile design. This was made possible by having the engine crankshaft in line with and at the same height as the transmission input shaft. There were several benefits of a low profile, not least of which was greatly improved visibility and better access for engine

White 2-60

USA

Specifications for 2-60 (1977)
Engine: Water-cooled four-cylinder
Bore x stroke: 3.9 x 4.3in (98 x 108mm)
Capacity: 211ci (3,292cc)
PTO power: 63hp @ 2,400rpm
Drawbar power: 40.4hp
Transmission: 8-speed
Speeds: 1.5-15.5mph (2.4-24.8km/h)
Fuel consumption: 14.4hp hr/gal
Weight: 5,160lb (2,322kg)

Although its first self-branded tractor was a big one, White wasn't about to abandon the lower end of the market. However, it was increasingly difficult to make lower-price tractors in the United States, and still make a profit. So White did what many others had and would do, and bolted its badge onto an imported tractor.

The new White 2-60 of 1976 was really a four-cylinder Fiat, sold initially in Oliver colours but soon changing to the new corporate silver colour scheme. In fact, there were two, the 47hp 2-50 and 63hp 2-60. Allis-Chalmers did the same at this time – its 5040 was built by Fiat. But neither of these Whites lasted very long, as the company soon turned to Iseki of Japan to fill the small end of its tractor range. After the 2-50 and 2-60 were dropped, four utility machines were offered, between 28 and 61hp, and all built by Iseki.

But the bigger two-wheel-drive tractors were still assembled in North America. The 2-105 for example, with its spacious glassy cab and

Above: Caterpillar power, Oliver transmission, White tractor.

servicing. Within a year, the 4-150 had been uprated into a 4-180, specifications for which are given above.

Above: The 2-60 was a rebadged Fiat.

turbocharged Perkins six-cylinder diesel of 354ci (5,522cc). It offered 105hp at the PTO, and eighteen forward gears. There was also a non-turbo 2-85. Top two-wheel-drive White in 1975 was the 2-150, this time using White's only six-cylinder engine of 585ci (9,126cc), and based on the old Minneapolis-Moline 1355.

White 2-70

USA

Specifications for 2-70 (1976)
Engine: Water-cooled six-cylinder
Bore x stroke: 3.875 x 4.0in (97 x 100mm)
Capacity: 283ci (4,415cc)
PTO power: 71hp @ 2,200rpm
Drawbar power: 57hp
Transmission: Eighteen-speed
Speeds: 2.1-16.6mph (3.4-26.6km/h)
Fuel consumption: 11.92hp hr/gal
Weight: 8,630lb (3,884kg)

What might look to the consumer (if they weren't particularly switched on) like an all-new range of White tractors in 1974-76 was actually a rebadging and repainting of existing parts. The 4-150 used many Oliver parts, the 2-60 was really a Fiat and the 2-150 a Minneapolis-Moline. And the same was true of the

2-70, introduced at the same time as the imported 2-50/2-60.

But the 2-70 wasn't imported, being based on the old Oliver 1655. It came with a 70hp gasoline engine of 265ci (4,134cc) and was in fact the last gasoline tractor offered by White. More relevant to most buyers now was the 283ci (4,415cc) direct injection diesel – both gasoline and diesel were six-cylinder units. There were three transmission options: Hydra-Power Drive, Over/Under Hydraul-Shift or Creeper Drive.

But as far as White was concerned, its tractor business wasn't doing well enough, a fact underlined by the increasingly tough trading conditions of the late 1970s. In 1980, it sold White Farm Equipment (including the Oliver, Cockshutt and M-M names) to the Texas Investment Group, a group of financiers that had no background in the tractor industry. Under Texas, the White group was renamed WFE, though tractors continued to be marketed under the White name.

Below: Looks familiar? The White 2-70 was a repainted Oliver 1655.

White 4-225

USA

Specifications for 4-210 (1979)
Engine: Water-cooled V8 diesel
Bore x stroke: 4.5 x 5.0in (113 x 125mm)
Capacity: 636ci (9,922cc)
PTO power: 159hp @ 2,800rpm
Drawbar power: 127hp
Transmission: Eighteen-speed
Speeds: 2.2-18.3mph (3.5-29.3km/h)
Fuel consumption: 12.34hp hr/gal
Weight: 23,735lb (10,681kg)

Under the new regime of Texas Investment Group ownership little actually changed. Five new Isekis were brought in to fill out the lower end of the range, from 28 to 75hp. Otherwise there were few changes apart from the launch of two new four-wheel-drive tractors in 1984, the 225hp 4-225 pictured here and 270hp 4-270.

Today, no tractors this powerful are sold with the White badge, though the closest is the 6215 Powershift. This four-wheel-drive tractor has 215hp at the PTO, courtesy of a six-cylinder Cummins C series diesel of 504ci, which is turbocharged and inter-cooled. Closed centre hydraulics have a 2,750psi relief valve, allowing multi-function hydraulic operation. Lower down the range, the 8310 and 8410 Fieldmaster tractors offer 125 and 145 PTO hp respectively. They are, according to White, 'born tough row-crop tractors with all the style and vitality of a new generation.'

But White had plenty to go through before reaching the 21st century. Only a year after the 4-225 was launched, Texas Investment Group gave up on White, and sold it to the Allied Products Corporation of Chicago. Allied already owned the New Idea name, and in 1987 combined the two into White-New Idea.

Below: Still badged as White, but under new ownership.

White 60

USA

Specifications for White 60 (1989)
Engine: Water-cooled, four-cylinder
Power: 61hp
Transmission: Six-speed (eighteen-speed option)

White's latest owners seemed a bit more serious about investing in new tractors, with big changes announced for the US-built end of the range in 1986. Four new machines were unveiled, ranging from 94 to 188 hp – the old Field Boss name was dropped, and the new machines were known as plain Whites 100, 120, 140 and 160. All had an 18-speed transmission, with the option of front-wheel-assist. Front-wheel-assist was an increasingly popular option in the 1980s, offering as it did some of the traction benefits of four-wheel-drive without the cost and complication of a full mechanical system. Power units for all the domestically produced tractors now came from CDC, a joint venture between Case and Cummins.

For about a decade, it had seemed accepted that nobody could profitably make small/mid-size tractors in North America any more. Then in 1989 White announced the new 60 and 80, with 61hp and 81hp respectively. These row-crop tractors were the smallest tractors sold in America that were actually made there as well. A basic six-speed transmission was standard, with a three-speed powershift option.

Right: The White 'American' underlined the home-built credentials.

White 80

USA

Specifications for White: 80 (1989)
Engine: Water-cooled, four-cylinder diesel
Capacity: 250ci (3.9 litres)
Power: 81hp
Transmission: Six-speed (eighteen-speed option)

Hoping to capitalise on the fact that the 60 and 80 were 'domestics' they were patriotically named the American 60 and 80. They even came in a choice of colours: Oliver green, Cockshutt red, Minneapolis-Moline gold or White silver. Someone at company HQ had cottoned on to the fact that these illustrious names might have some value. This ploy didn't last long however, and now all White tractors come in White silver only. Whatever, the 80's 3.9 litre Cummins engine made it the first sub-90hp tractor built in America for a decade.

The modern, 2002 equivalents of the White 60 and 80 come in two forms. The 6045/6065 general purpose tractors start with a 45hp three-cylinder diesel, without turbocharger. They come with two- or four-wheel-drive and a 12-speed Synchro-Reverse transmission, which allows quick direction changes in any gear. There's also the 6090, a high-clearance tractor with 80 PTO hp and syncromesh transmission giving 20 speeds in both forward and reverse. Maximum clearance is 27.2 inches,with 21.7 inches under the drawbar. Half a class up, the 6410/6510 are 70 and 85hp general purpose machines while the 6710/6810 offer 95 and 100hp, 'engineered from the ground up to deliver superior performance.'

Right: The White 60 and 80 survived into the 1990s.

WHITE
American
60

WHITE
80

White 6105

USA

Specifications for White 6195 (2002)
Engine: Water-cooled, six-cylinder turbo-diesel
PTO power: 200hp
Transmission: Eighteen-speed F / nine-speed R

In 1993, White had yet another change of ownership, and nine years later it looks as though the name may have finally found a stable corporate haven. AGCO, just as White had been 20 years before, was a new name to the tractor business. It had been formed in 1990 as a management buy-out of the American end of Deutz-Allis. AGCO began assembling a range of tractors using Deutz and Allis parts, importing smaller machines from SAME of Italy. And in a busy 1993, it added the North American arm of Massey-Ferguson to the stable as well as White, while a couple of years later McConnell (makers of articulated

x4 tractors) was bought as well. In a few short years, AGCO had become merica's biggest tractor maker. And just like White in the 1970s, AGCO sells everal different brands of tractor which share many common components.

From the mid-1990s (6215 apart) the top of the AGCO-White range onsisted of the big 6100-series, a range of six-cylinder turbo diesel tractors. he 6105 pictured here was the lead-in model, though that was later uprated ito the 6125, which uses a 5.9 litre Cummins – White has remained faithful to s long-time engine supplier. For 2002, the 6175 and 6195 Powershift were sted, giving 124hp and 200hp at the PTO respectively. The Powershift is now lectronically controlled by a single lever, which White says gives the 18 orward and 9 reverse gears more range than any other available transmission.

3elow: White is now part of the AGCO empire.

Zetor

Specifications for Zetor 7321 (2002)
Engine: Water-cooled, four-cylinder turbo-diesel
Power: 78hp @ 2,200rpm
Transmission: Ten-speed
Speeds: 1.1-18.4mph (1.8-29.4km/h)

Zetor of Czechoslovakia began life in November 1945, when the first prot[illegible] tractor was completed. Skoda and others had produced small numbe[illegible] tractors before the war, but the Zetor was intended for mass-production. [illegible] a couple of years, it accounted for 60 per cent of the country's entire tr[illegible] production. This little 25hp twin-cylinder diesel tractor was exported al[illegible] from the start, and was soon joined by a higher powered 30. A four-cyl[illegible] 42hp model arrived in 1954, in both wheeled and crawler form.

In the early '60s came a new unified range, with two- three- and [illegible] cylinder machines all sharing many compoentns. And the factory see[illegible] determined to keep up with Western tractor technology. The unified range [illegible] Zetor-matic hydraulic lifts and ten-speed transmissions, while four-wheel-[illegible] versions followed and Zetors were fitted with quiet safety cabs before [illegible] major European manufacturers. Higher powered Zetors, such as the [illegible] cylinder turbocharged 12011, followed.

By the late 1970s, Zetor had built half a million tractors, having acqui[illegible] good reputation in Western Europe for combining a relatively adva[illegible] specification and low price. The 7540 pictured here uses Zetor's own [illegible] cylinder diesel.

Below: Zetor sells in Europe as a reliable, good-value tractor.